MINISTRY OF AGRICULTURE, FISHERIES AND FOOD

Technical Bulletin 15

NITROGEN AND SOIL ORGANIC MATTER

Proceedings of a Conference
organized by the Soil Scientists of the
National Agricultural Advisory Service
October 22nd-23rd, 1964

LONDON
HER MAJESTY'S STATIONERY OFFICE
1969

First published 1969

Published by
Her Majesty's Stationery Office

To be purchased from
49 High Holborn, London W.C.1
13A Castle Street, Edinburgh 2
109 St. Mary Street, Cardiff CF1 1JW
Brazennose Street, Manchester 2
50 Fairfax Street, Bristol BS1 3DE
258 Broad Street, Birmingham 1
7 Linenhall Street, Belfast BT2 8AY
or through any bookseller

Printed in England for Her Majesty's Stationery Office, by Elliott Bros. & Yeoman Ltd., Liverpool

SBN 11 240915 6

Contents

Introduction

FOLLOWING the success of the conferences on Soil Phosphorus held in 1962[1] and Soil Potassium and Magnesium held in 1963[2] a similar conference on Nitrogen and Soil Organic Matter was held in 1964. The conference was organized by the Soil Scientists of the National Agricultural Advisory Service, with advisers and research workers from the United Kingdom and the Republic of Ireland participating.

Papers were presented on research and advisory aspects of nitrogen, its relationship with soil organic matter, analytical techniques which may be employed to estimate nitrogen available to crops and the influence of soil and environment on nitrogen residues.

The complete papers and accounts of the discussions following the papers are presented in this publication.

The outcome of the conference has been that advice on the nitrogen manuring of agricultural and horticultural crops in England and Wales has been rationalized and recommendations on the use of nitrogen have been assembled with those on phosphate and potash, and published as an Advisory Paper[3].

N. H. PIZER
Senior Science Adviser
National Agricultural Advisory Service

Ministry of Agriculture, Fisheries and Food
April 1968

---

1. Technical Bulletin No. 13 Soil Phosphorus—1965—H.M.S.O. London.
2. Technical Bulletin No. 14 Soil Potassium and Magnesium—1967—H.M.S.O. London.
3. Advisory Paper No. 4—Fertilizer Recommendations for Agricultural and Horticultural Crops—1967—M.A.F.F. London.

N.A.A.S. Soil Scientists and Visitors attending Conference

Chairman—Mr. F. Hanley, University of Cambridge

Dr. N. H. Pizer, Senior Science Adviser, National Agricultural Advisory Service, Headquarters

N.A.A.S. Soil Scientists:

Mr. F. C. Archer
Dr. T. Batey
Mr. C. Berryman
Mr. J. W. Blood
Mr. T. H. Caldwell
Mr. R. R. Charlesworth
Mr. C. G. Chumbley
Dr. D. B. Davies
Mr. G. R. Davies
Mr. H. T. Davies
Mr. W. Dermott
Mr. D. J. Eagle
Mr. G. H. A. Edwards
Mr. C. Evans
Mr. K. Farrar
Mr. C. S. Harper
Mr. M. F. Harrod
Mr. T. G. Heafield
Mr. D. Hewgill
Mr. L. J. Hooper
Mr. E. Jones
Mr. J. Langmead
Dr. R. C. Little
Mr. P. Needham
Mr. J. B. E. Patterson
Mr. E. Roberts
Dr. T. H. Rose
Mr. R. D. Russell
Mr. W. O. Sharp
Dr. K. Shaw
Dr. C. W. R. Smith
Mr. R. W. Swain
Mr. R. Thomas
Mr. G. A. Wadsworth
Mr. W. M. Waller
Dr. J. E. Watkin
Mr. J. Webber
Mr. B. Wilkinson
Mr. J. H. Williams

Visitors:

Dr. G. Anderson, Macaulay Institute for Soil Research, Aberdeen
Mr. E. R. Armitage, Jealotts Hill Research Station, Bracknell, Berks.
Mr. H. L. Back, Grassland Research Institute, Hurley, Nr. Maidenhead, Berks.
Dr. C. Bould, Long Ashton Research Station, Bristol
Dr. D. A. Boyd, Rothamsted Experimental Station, Harpenden, Herts.
Mr. J. Brockman, North Wyke Experimental Station, North Tawton, Devon
Mr. J. Brogan, Johnstown Castle Agricultural College, Wexford
Mr. D. Browne, Dairying Research Centre, Moorepark Fermoy, Co. Cork
Mr. A. C. Bunt, Glasshouse Crops Research Institute, Littlehampton, Sussex
Dr. C. R. Clement, Grassland Research Institute, Hurley, Nr. Maidenhead, Berks.
Dr. G. W. Cooke, Rothamsted Experimental Station, Harpenden, Herts.
Miss A. W. Cooper, Bacteriology Dept., N.A.A.S., Newcastle
Mr. D. W. Cowling, Grassland Research Institute, Hurley, Nr. Maidenhead, Berks. (Deputizing for Mr. J. O. Green)
Mr. P. Crooks, School of Agriculture, Edinburgh
Mr. R. I. Davies, University College of North Wales, Bangor

Dr. P. C. DeKock, Macaulay Institute for Soil Research, Aberdeen
Mr. J. Devine, Levington Research Station, Ipswich, Suffolk
Mr. J. W. Egdell, Bacteriology Dept., N.A.A.S., Bristol
Dr. J. K. R. Gasser, Rothamsted Experimental Station, Harpenden, Herts.
Dr. D. J. Greenwood, National Vegetable Research Station, Wellesbourne, Warwicks.
Dr. D. W. P. Greenham, East Malling Research Station, East Malling, Kent
Dr. G. ap Griffiths, Welsh Plant Breeding Station, Plas Gogerddan, Nr. Aberystwyth
Dr. E. J. Hewitt, Long Ashton Research Station, Bristol
Mr. F. Haworth, National Vegetable Research Station, Wellesbourne
Mr. D. G. Hill-Cottingham, Long Ashton Research Station, Bristol
Mr. M. R. J. Holmes, Levington Research Station, Ipswich
Dr. D. S. Jenkinson, Rothamsted Experimental Station, Harpenden, Herts.
Mr. A. E. Johnston, Rothamsted Experimental Station, Harpenden, Herts.
Mr. E. A. Kirkby, The University, Leeds
Mr. P. Last, Brooms Barn Experimental Station, Suffolk
Mr. W. J. Lessells, Rothamsted Experimental Station, Harpenden, Herts.
Mr. A. J. Low, Jealott's Hill Research Station, Bracknell, Berks.
Mr. J. S. V. McAllister, Queens University, Belfast
Dr. T. Z. Nowakowsky, Rothamsted Experimental Station, Harpenden, Herts.
Mr. D. Reid, Department of Grass and Dairy Husbandry, Hannah Dairy Research Institute, Kirkhill, Ayr
Prof. E. W. Russell, Dept. of Soil Science, Reading University, Reading, Berks.
Mr. K. Simpson, School of Agriculture, Edinburgh
Dr. P. B. Tinker, Brooms Barn Experimental Station, Suffolk
Dr. J. Tinsley, University of Aberdeen, Aberdeen
Mr. W. N. Townsend, Department of Agricultural Chemistry, The University, Leeds
Mr. R. G. Warren, Rothamsted Experimental Station, Harpenden, Herts.
Dr. G. C. White, East Malling Research Station, East Malling, Kent
Mr. F. V. Widdowson, Rothamsted Experimental Station, Harpenden, Herts.
Mr. W. S. Wilson, Institute of Agriculture, Writtle, Essex
Dr. G. W. Winsor, Glasshouse Crops Research Institute, Littlehampton, Sussex

---

Joint editors of Conference Technical Bulletin

L. J. Hooper, N.A.A.S., Coley Park, Reading, Berks.
D. J. Eagle, N.A.A.S., Brooklands Avenue, Cambridge

Chairman's opening remarks

The Chairman extended a welcome to all attending the conference. He said that discussion was the main purpose of the conference and hoped that all would join in the discussion of the papers presented. Many of the points which emerged from the discussions at the two previous conferences, 'Soil phosphorus' and 'Soil potassium and magnesium' had been followed up by N.A.A.S. Soil Scientists and it was hoped that equally useful pointers would emerge at the present conference. The delegates were reminded that the conference would divide into syndicates, to discuss certain topics; the syndicates, would later report to the full conference. Speakers were asked to bear in mind that the main theme of the conference was the relationship of 'Nitrogen and Soil Organic Matter' to advisory work.

The metabolism of nitrogen in plants

By P. C. DeKOCK, M.Sc., D.Phil.
Macaulay Institute for Soil Research, Aberdeen

THE fuller understanding of the metabolism of nitrates and other nitrogenous compounds in plants is very desirable since large amounts of nitrogenous fertilizers are used in agriculture and may represent a major item of farm costs. Inefficient utilization by crops entails needless waste. On the other hand high concentrations of nitrate in plant tissues may have undesirable effects such as the production of poisonous fumes during silage making, and nitrate toxicity in animals.

Since the ammonium ion is rapidly converted into nitrate in the soil by nitrifying bacteria, plants absorb it in this form and must reduce nitrate back to ammonium before it can be incorporated into amino acids. The stages in the reduction may be represented as follows:

$$\underset{\text{nitrate}}{NO_3^-} \longrightarrow \underset{\text{nitrite}}{NO_2^-} \longrightarrow \underset{\text{hyponitrite}}{(NOH)_2^-} \longrightarrow \underset{\text{hydroxylamine}}{NH_2OH^+} \longrightarrow \underset{\text{ammonium}}{NH_4^+}$$

The enzymes involved in these reductive stages have been studied in some detail.

Nitrate reductase

This enzyme which reduces nitrate to nitrite, was first isolated from the fungus *Neurospora* by Evans and Nason[3] and subsequently demonstrated in higher plants[4,14]. The reaction is essentially

$$NO_3^- + DPNH + H^+ \longrightarrow NO_2^- + DPN^+ + H_2O$$

The enzyme from green plants appears to use diphosphopyridine nucleotide (DPNH) for reduction but triphosphopyridine nucleotide (TPNH) is required for the enzyme prepared from soybean nodules.

By purification some seventy fold, the enzyme has been shown to be a flavoprotein with a flavine-adenine dinucleotide (FAD) as prosthetic group and containing the element molybdenum[13]. The electron-transporting chain can thus be represented as:

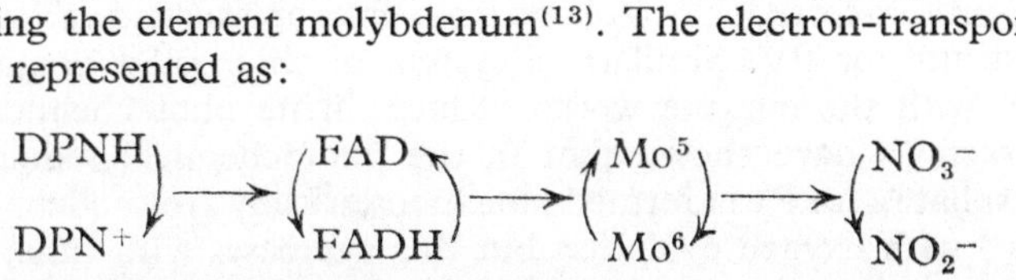

Evidence in favour of the above sequence has been provided by Nicholas and Nason[12]. Molybdenum can be removed from the enzyme by dialysis, whereupon the molybdenum-free enzyme will still reduce FAD but not reduce nitrate. Only when molybdate is added, is the reduction of nitrate possible. Plants deficient in molybdenum have long been known to accumulate nitrate[17].

Hageman and co-workers[5,6] have studied the occurrence of nitrate reductase in plants. The sap pH optimum for most plants seems to be in the region of 5–6 but for soybean it is somewhat higher. It appears to be a labile enzyme, its activity declining rapidly in the dark and reappearing in the light. This activity is inversely correlated with the content of nitrate in plant tissues which normally declines during the day and rises during the hours of darkness. Shading is therefore also an important factor in accumulation of nitrate in plants; close planting and consequent overcrowding may give rise to high nitrate levels. The enzyme is only found when nitrate is present in the plant and appears to be inhibited by the ammonium ion.

Seasonal variation in the activity of the enzyme in maize plants has been observed, activity being greatest at the time of most rapid growth and then declining but reaching a second peak at the time of flower formation. Inadequate water supply causes a decline in enzyme activity which is rapidly restored when the water balance is re-established. Of great interest is the fact that the weedkiller 2,4 dichlorophenoxy-acetic acid (2,4-D) reduces the activity of the enzyme in dicotyledonous plants but in monocotyledons the activity is enhanced.

Nitrate reductase

This enzyme has been studied in some detail in *Neurospora* by Medina and Nicholas[10] and found to be a flavoprotein requiring iron, copper, phosphate and pyridoxin for its activity. The reaction seems to involve a phosphorylation as it is inhibited by 2,4 dinitrophenol. It is also inhibited by hyponitrite, presumably an intermediate in the reaction.

Hageman, Cresswell and Hewitt[5] have investigated the reduction of nitrate and nitrite in plants. Nitrate reduction is usually measured *in vitro* by the nitrite formed, but if nitrite is infiltrated into a leaf, its disappearance is rapid. This indicates that the accumulation of nitrite during nitrate reduction is due to the absence of the necessary factors for further reduction to take place. Thus if the dye, benzyl viologen (BV) is used instead of DPNH, nitrite does not accumulate. A study of the system showed nitrite reduction to be dependent upon the relative amounts of oxidised and reduced dye. Rates of 6,000 μMoles nitrite/hr/g fresh weight were recorded and recovery of ammonia varied from 85–100 per cent. Hydroxylamine was reduced at about half the above rate and gave lower recoveries of ammonia, presumably due to oxime formation. TPNH would reduce BV and if glucose-6-phosphate and its dehydrogenase enzyme were added, nitrite reduction would proceed in an open tube.

The iron-protein ferredoxin, isolated from chloroplasts and various organisms would substitute for BV. Similarly a system of chloroplast grana, ferredoxin and TPNH with the enzyme would reduce nitrite photochemically. Losada and co-workers[15] have shown that in the photochemical reduction, adenosine triphosphate (ATP) is formed stoichiometrically from adenosine diphosphate (ADP) with oxygen evolution but this decreases with time, presumably due to uncoupling by the ammonia formed. The equation is given as:

$$2\,NO_2^- + 6\,ADP + 6\,H_2O + 6\,H_3PO_4 + 2\,H^+ \xrightarrow{\text{light}} 2\,NH_3 + 6\,ATP + 3O_2 + 10\,H_2O$$

The relationship between moles nitrite reduced, moles ammonia formed, moles ATP generated and atoms of oxygen evolved is thus 1 : 1 : 3 : 3.

A certain specificity in the type of ferredoxin used was observed, spinach ferredoxin being more efficient than *Clostridium* ferredoxin in the light reduction of nitrite, but less efficient in the dark reduction with methyl viologen and TPNH. Ferredoxin could thus serve as electron acceptor between hydrogen gas, the photochemical reaction or TPN and the appropriate enzyme, as shown in the following diagram:

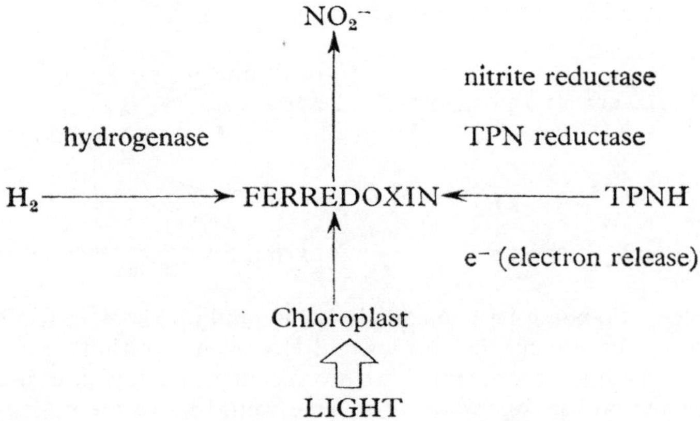

The metabolism of ammonia

The ammonium ion is rapidly absorbed by plants and incorporated into amino acids, so that its concentration in most plant tissues is not very great. It has been found as a salt of organic acids in acidophilous plants. Generally quite small concentrations are toxic[16] and the mechanism of detoxification is formation of an amide with glutamic acid. Isotopically labelled ammonia fed to a plant is quickly detected in glutamic acid. The mechanism of incorporation is considered to be reaction with α-ketoglutaric acid to form the corresponding amino acid which on reduction yields glutamic acid. The formation of the amide involves phosphorylation:

$$\text{Glutamic acid} + NH_3 + \text{ATP} \longrightarrow \text{glutamine} + \text{ADP} + \text{P (inorganic)}$$

Similar phosphorylation reactions are involved in the synthesis of peptides and proteins from amino acids.

Urea

Urea is very easily absorbed by plants and is then thought to be hydrolysed to ammonium carbonate by the enzyme urease:

$$CO(NH_2)_2 + 2\,H_2O \longrightarrow (NH_4)_2\,CO_3$$

It is, however, doubtful whether urea is completely hydrolysed to ammonium carbonate and there is evidence that carbamyl phosphate is produced:

$$\underset{\text{urea}}{CO(NH_2)_2} + H_3PO_4 \rightleftharpoons \underset{\text{carbamyl phosphate}}{H_2N - \overset{\overset{O}{\|}}{C} - O\overset{\overset{O}{\|}}{P}(OH) - OH} + NH_3$$

Carbamyl phosphate is an energy-rich compound and an intermediate in pyrimidine synthesis. Ammonia itself may react with CO_2 and ATP to produce carbamyl phosphate:

$$2\,ATP + NH_3 + CO_2 \longrightarrow H_2NCO\,.\,OPO_3H + 2\,ADP + H_2PO_4^-$$

The following scheme has been given:

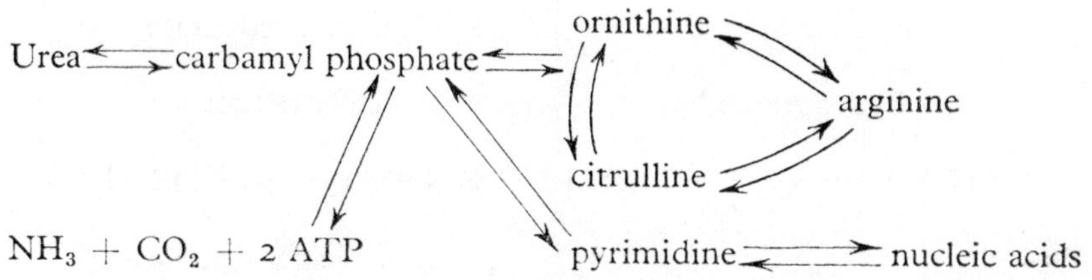

Some plants do not appear to contain urease and urea may in fact be utilized by such plants by the classical Krebs and Henseleit Ornithine cycle operating in reverse. Labelling experiments have given evidence in favour of the ornithine cycle in plants and its operation in reverse would make the reaction energy-releasing. The synthesis of purines would probably require the synthesis of urea. Allantoin and allantoic acid occur in quantity in plants and may be the major translocatory form of nitrogen(11). Hence urea sprays may lead to synthesis of allantoin and a decrease of glycine in accordance with the idea that allantoin synthesis takes place from urea and glyoxylic acid. However, Krupka and Towers(8) doubt that this is the case and propose the following scheme:

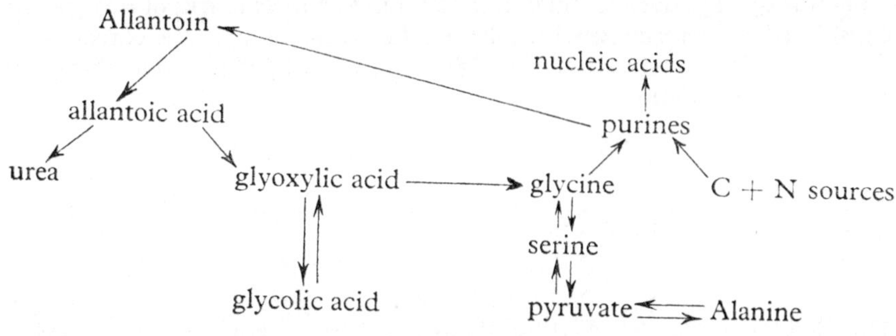

The field of nitrogen nutrition is vast and only passing reference can be made to points of salient interest here. There is still much to learn. Bollard(1) could detect no nitrate in the tracheal sap of the apple tree, the interference being that all nitrate is completely reduced in the roots by a dark reaction. Using isotopically labelled nitrogen, Delwiche(2) could find no evidence that ammonia was oxidized to nitrate in plants so that nitrate reduction does not appear to be reversible. However, it has recently been reported that this may happen in the green alga *Chlorella* and all the enzymes necessary for the oxidation have been identified(18). Also Vickery and co-workers(19) observed the nitrate content of isolated tobacco and other leaves to increase and concluded that nitrate is in equilibrium with other nitrogenous constitutents of the leaf.

There are also certain plants containing aromatic nitro-compounds. It would be interesting to know if the formation of these compounds is suppressed under conditions of ammonium fertilization.

References

1. Bollard, E. G. (1959). *Symp. Soc. Expl. Biol.* **13**.
2. Delwiche, C. (1951). *J. Biol. Chem.* **189**, 167.
3. Evans, H. J. and Nason, A. (1952). *Arch. Biochem. Biophys.* **39**, 234.
4. Evans, H. J. and Nason, A. (1953). *Plant physiology* **28**, 233.
5. Hageman, R. H., Cresswell, C. F. and Hewitt, E. J. (1962). *Nature* **193**, 247.
6. Hageman, R. H., Beevers, L. and Hewitt, E. J. (1964). Comm. 10th Int. Bot. Congr., Edinburgh.
7. Hewitt, E. J. and Jones, E. W. (1947). *J. Pomol. Hort. Sci.* **23**, 254.
8. Krupka, R. M. and Towers, G. H. N. (1959). *Canad. J. Bot.* **37**, 539.
9. Losada, M., Paneque, A., Ramarez, J. M. and Del Campo, F. F. (1963). *Biochem. Biophys. Res. Comm.* **10**, 298.
10. Medina, A. and Nicholas, D. J. D. (1957). *Biochem. Biophys. Acta* **25**, 138.
11. Mothes, K. and Engelbrecht, L. (1954). *Flora* **141**, 356.
12. Nicholas, D. J. D. and Nason, A. (1954). *Arch. Biochem. Biophys.* **51**, 310.
13. Nicholas, D. J. D., Nason, A. and McElroy, W. D. (1954). *J. Biol. Chem.* **207**, 341.
14. Nason, A. (1956). *Inorganic nitrogen metabolism of plants*. McElroy and Glass, Baltimore.
15. Paneque, A., Ramirez, J. M., Del Campo, F. F. and Losada, M. (1964). *J. Biol. Chem.* **239**, 1737.
16. Warren, K. S. (1962). *Nature*, **195**, 47.
17. Wilson, P. W. and Waring, E. J. (1948). *J. Austr. Inst. Agr. Sci.* **14**, 141.
18. Yamafugi, K. and Osajima, Y. (1963). *Enzymologia* **26**, 75.
19. Vickery, H. B., Pucher, G. W., Wakeman, A. J. and Leavenworth, C. S. (1933). Carnegie Inst. Washington Publication No. 445.

Discussion

Dr. D. W. P. Greenham said he wondered how the light dependence of the activity of nitrate reductase could be reconciled with the fact that nitrate is reduced in the rootlets of apple trees. *Dr. DeKock* replied that this observation did not agree with work that had been done on other crops such as the cucumber and he thought that the apple would be a good subject for further research.

Dr. G. ap Griffiths said that in spite of the experimental proof of the fact that the action of nitrate reductase in the leaf was light dependent and that shading of leaves led to a higher leaf nitrate content it should not be overlooked that the main reason for accumulation of nitrate in the leaves was the presence of an excess of nitrate in the plant.

Dr. R. C. Little asked how much nitrite plants could tolerate, as it had been suspected that nitrite was involved in post-steaming injury to crops growing in glasshouse soils. *Dr. DeKock* replied that work was at an early stage but plants could utilize nitrite in nutrient solution at a concentration of 2/5 milli equiv. per litre without harmful effects.

Dr. N. H. Pizer commented that in a greenhouse experiment, nitrite had been applied to sugar beet which grew satisfactorily except that the leaves were pale. The leaves were analysed and the nitrite content was high.

Mr. L. J. Hooper stated that in Wiltshire it had been found that cereal leaves contained high nitrate levels in dull wet seasons and asked if the light effect on reduction of nitrate applied to each stage of the reduction to ammonia. *Dr. DeKock* replied that according to the literature the nitrate to nitrite stage was the rate limiting stage.

Dr. T. Z. Nowakowsky described a greenhouse experiment in which Italian ryegrass was grown under low light intensity at three soil temperatures—11 °C, 19 °C and 28 °C. With increasing soil temperature the amount of nitrate in the grass increased very considerably and he ascribed this to the depletion of the carbohydrates in the plant brought about by the increased rate of respiration.

Dr. DeKock asked if anyone present had seen a silo emitting red fumes and whether anyone had any knowledge of stock losses due to production of nitrite in silage.

Dr. N. H. Pizer commented that he had read the American reports of production of nitrogen oxides by ensiled grain but investigations had revealed no evidence of nitrite formation in silos in this country and attempts to produce it experimentally had failed.

Dr. G. ap Griffiths drew attention to a conflict between reports in the U.S.A. that nitrate was toxic to stock and experiments at the Welsh Plant Breeding Station which had failed to show harmful effects due to nitrate. All reports of nitrate toxicity in the U.S.A. were associated with mouldy hay and experiments at the Welsh Plant Breeding Station had shown that nitrate could be reduced to nitrite during the deterioration of the hay. He suggested that the so-called nitrate poisoning was really due to nitrite formed in the hay in this way.

Mr. Kirkby said that a case of poisoning of children in Germany had been shown to be due to high nitrate levels in spinach.

Uptake by plants of various forms of nitrogen and effects on plant composition

By P. C. DeKOCK, M.Sc., D.Phil. and
E. A. KIRKBY*, B.Sc. (Dunelm), A.R.I.C.
Macaulay Institute for Soil Research, Aberdeen

OF the forms of nitrogen available in agriculture, the ammonium and nitrate ions are the most used. The fact that ammonia is rapidly nitrified in most soils has tended to obscure the different physiological effects of these two ions on plant growth and composition. However, differences in plant growth are frequently observed under field conditions between these forms of nitrogen and claims have been made that certain plants can only tolerate the one or the other form of nitrogen. Damage to field crops from fertilizers containing a high proportion of ammonia has frequently been reported. Conversely it has been found that certain plants may contain high concentrations of nitrate in their tissues and it has been inferred that the plant is unable to utilize this form of nitrogen.

Undoubtedly a considerable amount of the confusion which exists on this topic is due to the fact that absorption of the ammonium ion from its salt leads to increased acidity in the growth medium whereas absorption of the nitrate ion results in decreased acidity. These changes in pH affect the availability of the other nutrient elements, especially iron, so that the chlorosis or poor growth observed is only indirectly due to the form of nitrogen applied. Karim and Vlamis[16] have shown that rice can utilize nitrate quite readily as long as the pH of the medium is maintained at about 4. Similarly Sideris and Krauss[25] have shown that the two forms of nitrogen are equally effective in promoting the growth of pineapples provided the pH is maintained at 5.

Although the technique of nutrient culture has lessened the difficulties encountered in the comparison of these two forms of nitrogen, pH changes are still liable to take place as nitrogen is required by the plant in much greater quantities than the other elements. Such difficulties can be overcome by growing plants in large volumes of solution thereby minimizing pH changes[8]. It is also evident that since ammonium and nitrate are respectively a cation and an anion, variation in their concentration in the nutrient solution invariably involves the alteration of the anionic and cationic composition of the growth medium so that it is practically impossible to obtain conditions under which only one factor in the nutrient medium is varied.

However, commencing with the nutrient culture studies of Clark[4] and Arnon[1] there is now conclusive evidence to show that plants grow differently when supplied with the two forms of nitrogen and that their cationic and anionic composition is altered. Clark found tomato plants to grow more luxuriantly in solutions containing nitrate; they had a more robust appearance and broader leaves, even though the colour of the leaf was somewhat paler. A retardation of the growth of roots in the ammonium-fed plants was also

*Present address: Department of Agriculture, The University, Leeds.

observed. We have found similar results in the growth of tomato, mustard, buckwheat and rye. Evers[13] has also found this for trees of various species.

Although there seems to be no consistent difference between the percentage dry weight of the plants, the percentage ash in NH_3-fed plants is always considerably less than that of NO_3-fed plants. This increase in ash in the latter is largely accounted for, at any rate in dicotyledonous species, by an increase in the calcium content[3,5].

Most agricultural crops, in fact most plants, have a relatively constant sap pH which in many instances lies between 5 and 6. Hence the cationic composition is balanced or more than balanced by the sum of the anions (Fig. 1). Pierce and Appleman[23] first showed that the difference between the total cations and inorganic anions could be accounted for by the organic anions—the organic acids.

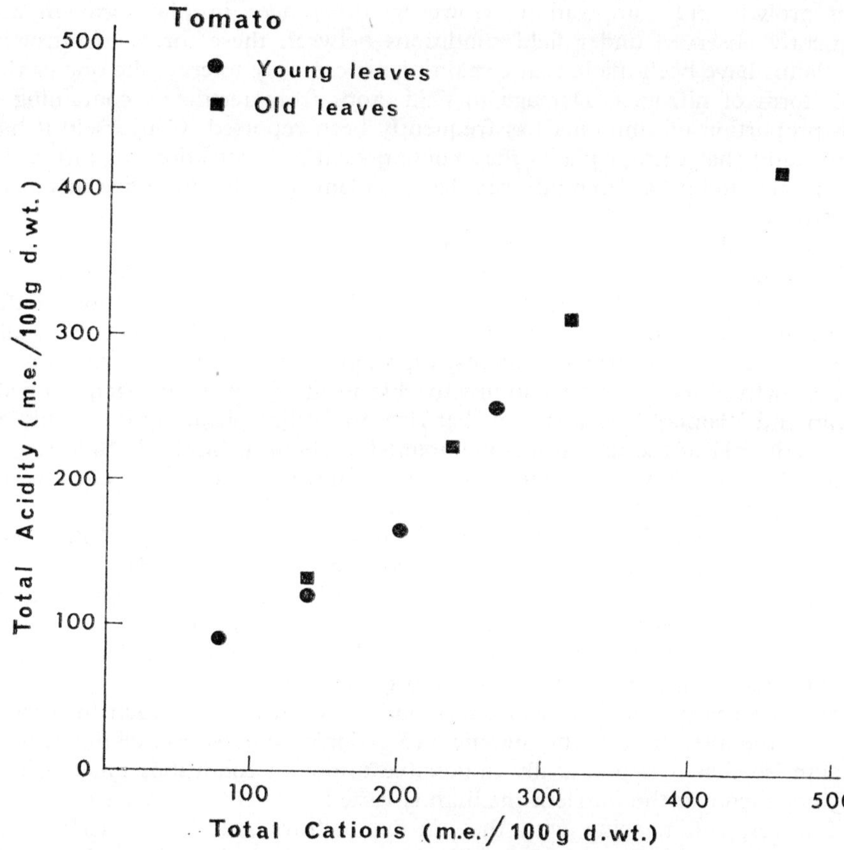

Fig. 1. Relation between total acidity and total cations in tomato leaves. (Calculated from Coic *et al.* 1962)

The early work of Ruhland showed that Begonia plants fed with nitrate contained more oxalic acid than those fed with ammonium. Clark similarly found more total organic acid in nitrate fed plants, of which malic and oxalic acids constituted the greater proportion. Moreover, in those plants which do not form oxalic acid, malic is greatly increased.

When plants are grown in various mixtures of ammonium and nitrate or changed from one nitrogen form to the other they show proportional response in cation and anion content (Fig. 2). Lycklama[19] has studied the rates of uptake of the ammonium and nitrate ions by plant roots. His results show that the ammonium ion is taken up about twice as fast as the nitrate ion and moreover that uptake of nitrate is depressed in the presence of ammonium. Syrett[27] has similarly shown that the alga Chlorella takes up the ammonium ion preferentially to nitrate when offered both together. The fact however that the change in mineral composition of the plant is proportional to the relative amounts of the two forms of nitrogen in the nutrient medium suggests that the rates of uptake by the plant of these forms of nitrogen are about equal and therefore the amounts taken up are dependent upon their external concentrations. This could mean that the plant is in fact in equilibrium with the external medium and that rate limiting reactions are not involved.

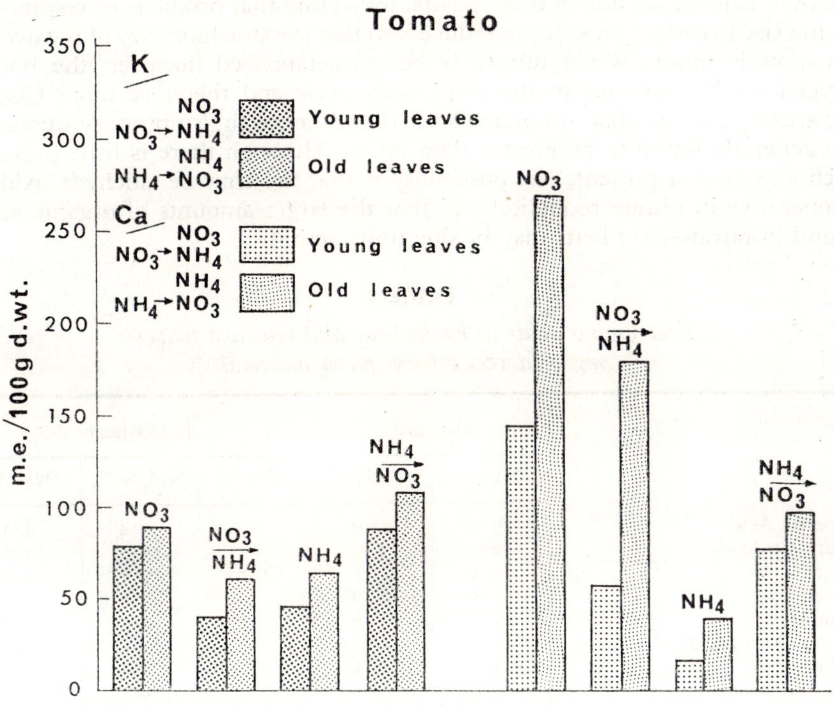

Fig. 2. Potassium and calcium content of tomato leaves as affected by nitrate or ammonium nutrition or passage from one to the other. (Data of Coic *et al.* 1962)

Concomitant absorption of cation and anion has been invoked to explain the differences in plant composition between these two forms of nitrogen. According to theories of Lundegärdh[18] absorption of an anion such as nitrate would cause a concomitant absorption of a cation to maintain electrical neutrality. Then if the nitrate ion is metabolized, the charge of the remaining cation would have to be compensated for by production of an organic acid. Dijkshoorn[11] has made extensive use of such postulates. However the demonstration that cations and anions can be taken up quite independently of one another[14] and that plants take up cations quite specifically, argues against such postulates.

Syrett[27] has attempted to explain the reduction of dicarboxylic organic acids under ammonium nutrition as due to the formation of amino acids from the keto acids in the tricarboxylic acid cycle. Ammonium-fed plants contain much greater quantities of free amino acids than nitrate-fed plants (Table 1) even though there is little or no difference between the nitrogen content of the two types. The isocitric shunt has been proposed as a means of recharging the tricarboxylic acid cycle when these acids are removed by other pathways. Millerd *et al.*[20] have demonstrated the operation of this pathway in *Oxalis pes-caprae* and it appears likely that this is the pathway by which oxalic acid is formed in the plant. So far, however, it is considered to be limited to fat-containing germinating seeds[2] and its operation in the plant is still an open question. It would appear therefore that the ammonium ion does cause depletion of the cycle. In accordance with this view is the fact that the respiratory quotient is less than one in these plants, indicating that products of respiration giving rise to carbon dioxide are reduced, so that the stoichiometry of the overall equation is upset. When nitrate is being metabolized however, the bound oxygen also contributes to the respiratory cycle and therefore more CO_2 is apparently evolved than oxygen absorbed and hence the respiratory quotients are generally found to be greater than unity. Although there is little proof of such a system at present, one possibility is that the enzyme aldehyde oxidase is operative in nitrate reduction and that the larger amounts of organic acids found in nitrate-fed plants may be due to its activity.

Table 1

Free amino acids in buckwheat and mustard leaves
mg acid/100 g fresh plant material

	Mustard		Buckwheat		
	NO_3^-	NH_4^+	NO_3^-	NH_4NO_3	NH_4^+
Aspartic Acid	29·5	30·9	Tr	3·4	4·11
Glumatic Acid	52·6	56·9	Tr	15·5	17·5
Serine	17·2	24·6	0·72	9·6	6·0
Threonine	7·5	8·7	—	Tr	Tr
Alanine	4·8	8·6	0·82	6·0	11·6
γAminobutryric Acid	3·2	9·5	—	—	—
Valine	5·3	7·3	0·34	5·6	4·1
Leucine/isoleucine	2·4	3·7	—	—	—
Phenylalanine	—	3·6	—	—	—
Lysine	—	2·9	—	—	—
Asparagine	—	2·1	—	—	—
Proline	48·8	72·8	—	—	—

Our earlier work on mineral nutrition of plants has shown that in plants grown under various experimental conditions, uptake of phosphorus is related to uptake of iron[9]. The phosphorus–iron ratio also bears a direct relationship to the potassium–calcium ratio. This is considered to be due to the poise in the tricarboxylic acid cycle since correlations have been obtained between the contents of malic (and/or oxalic) acid and the amounts of calcium[22].

These relationships are also apparent in plants supplied with ammonium and nitrate (Figs. 3, 4, 5) and this implies that the form of nitrogen affects metabolism in a way which will allow prediction of the uptake of other elements.

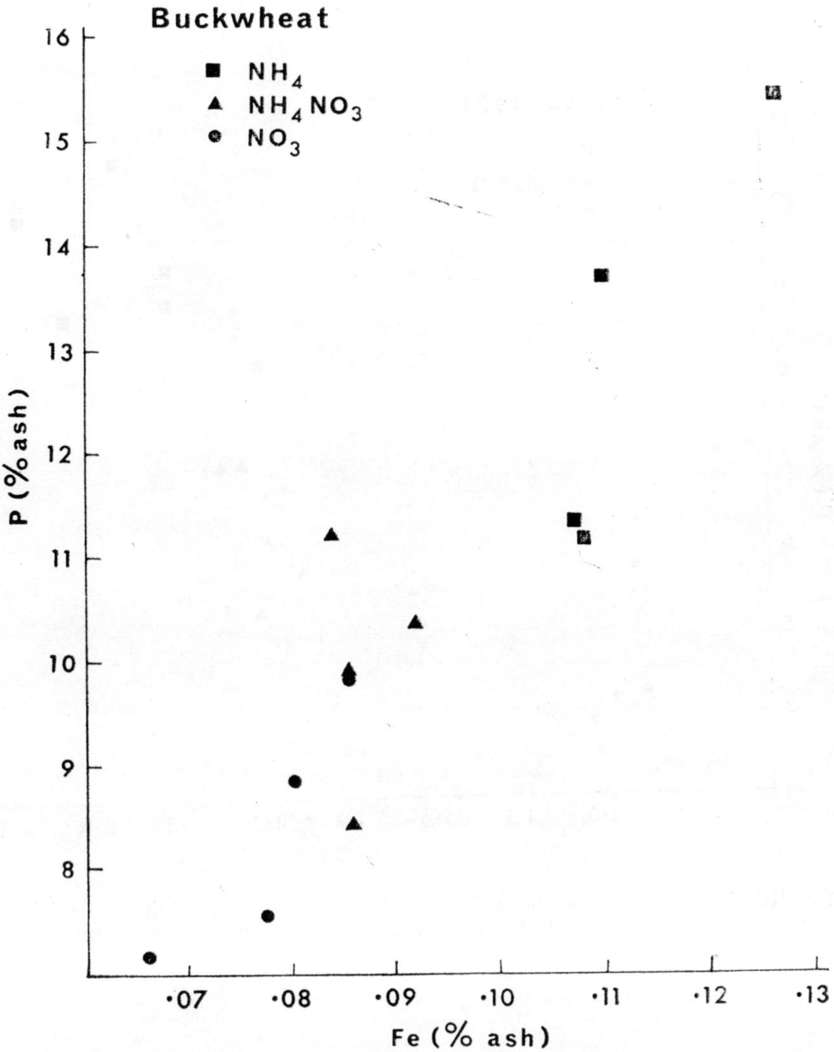

Fig. 3. Relation between total phosphorus and total iron in buckwheat leaves grown with nitrate, ammonium nitrate or ammonium. (Author's data)

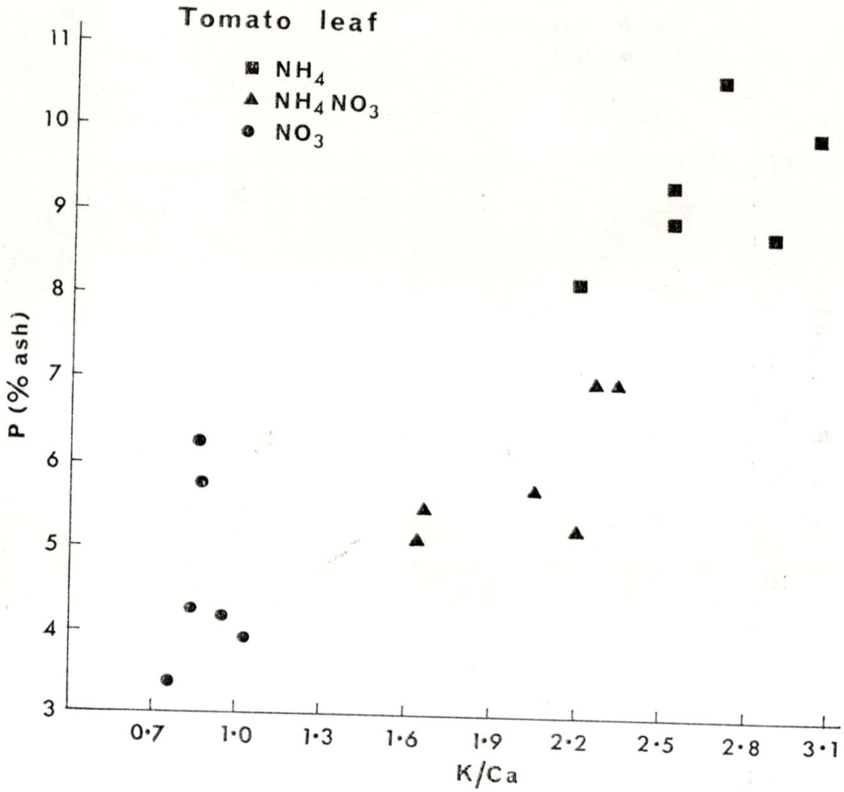

Fig. 4. Relation between phosphorus content and the ratio of potassium to calcium in tomato leaves grown with nitrate, ammonium nitrate or ammonium. (Author's data)

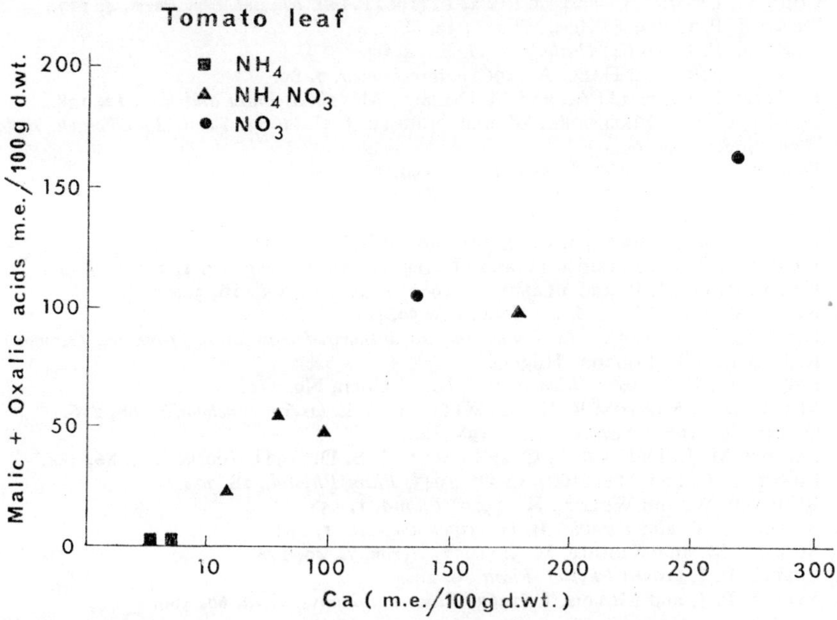

Fig. 5. Relation between calcium content and malic plus oxalic acids in tomato leaves. (Data of Coic *et al.* 1962)

The fact that soils in northerly areas are frequently too cold in the early dart of the year to allow nitrification to proceed coupled with the possibility of the introduction into agricultural practice of inhibitors of nitrification, lends more than academic interest to the effect of these forms of nitrogen on plants as it has been shown that the relative amounts of potassium and magnesium are important in the syndrome of hypomagnesia in cattle[17]. Also the rapid uptake of nitrate by plants frequently leads to high concentrations in the tissues and this has been cited as a cause of abortion in cattle[26]. Moreover nitrate reductase has been shown to be a labile enzyme[15] so that some understanding of the physiological interrelationships leading to nitrate accumulation and reduction is desirable.

Of the other forms of nitrogen used in agriculture, urea deserves some mention. It has recently been used with success as a post-harvest spray on apple trees[21] and spectacular increases in nitrogen content of apple leaves have been obtained overnight[12]. It would appear that urea is not analogous to ammonium in its action on plant composition. There is evidence that it is not necessarily hydrolyzed by the enzyme urease and may be incorporated into allantoin or other purines. Also some plants such as *Populus nigra* contain effective urease inhibitors in their tissues and therefore a study of their metabolism would be worthwhile.

References

1. Arnon, D. I. (1937) *Soil Sci.*, **44**, 91.
2. Carpenter, W. D. and Beevers, H. (1959) *Plant Physiol.*, **34**, 403.
3. Chouteau, J. (1960) *Ann. Physiol. veg. Paris*, **2**, 237.
4. Clark, H. E. (1936) *Plant Physiol.*, **11**, 5.

5. Coic, Y., Lesaint, C. and Le Roux, F. (1962) *Ann. Physiol. veg. Paris*, **4**, 117.
6. DeKock, P. C. (1958) *Agric. Progr.*, **33**, 88.
7. DeKock, P. C. (1964) *Outlook on Agric.*, **4**, 93.
8. DeKock, P. C. and Hall, A. (1962) *Agrochimica*, **7**, 89.
9. DeKock, P. C., Hall, A., and McDonald, M. (1960) *Plant and Soil*, **12**, 128.
10. DeWit, C. T., Dijkshoorn, W. and Noggle, J. C. (1963) *Versl. Landbouwk. onderz.* Wageningen, No. 69, 15.
11. Dijkshoorn, W. (1962) *Nature, Lond.*, **194**, 165.
12. East Malling Research Station Annual Report (1963) p.22.
13. Evers, F. H. (1964) *Z. Bot.*, **51**, 91.
14. Fried, M. and Noggle, J. C. (1958) *Plant Physiol.*, **33**, 139.
15. Hageman, R. H., Flesher, D. and Gitter, A. (1961) *Crop Sci.* **1**, 201.
16. Karim, A. Q. M. B. and Vlamis, J. (1962) *Plant and Soil*, **16**, 32.
17. Kemp, A. (1964) *C. R. Acad. Agric. Fr.*, **50**, 540.
18. Lundegärdh, H. (1951) *Leaf analysis: an authorised translation (from the German)* by R. L. Mitchell, (London: Hilger).
19. Lycklama, J. C. (1963) *Meded. Bot. Lab.* Leiden, No. 110.
20. Millerd, A., Morton, R. K. and Wells, J. R. E. (1963) *Biochem. J.*, **88**, 276.
21. Oland, K. (1963) *Nature, Lond.*, **198**, 1282.
22. Palmer, M. J., DeKock, P. C. and Bacon, J. S. D. (1963) *Biochem. J.*, **86**, 484.
23. Pierce, E. C. and Appleman, C. O. (1943) *Plant Physiol.*, **18**, 224.
24. Ruhland, W. and Wetzel, K. (1926) *Planta*, **1**, 558.
25. Sideris, C. P. and Krauss, B. H. (1937) *Growth*, **1**, 204.
26. Sund, J. M. and Wright, M. J. (1957) *Agron. J.*, **49**, 278.
27. Syrett, P. J. (1956) *Physiol. Plant.*, **9**, 28.
28. Syrett, P. J. and Morris, I. (1963) *Biochem. biophys. Acta*, **67**, 566.

Discussion

Mr. G. H. A. Edwards questioned *Mr. Kirkby*, on the synthesis of organic acids with special reference to the malic acid content of rhubarb. *Mr. Kirkby* replied that he had no specific knowledge of rhubarb but could be sure that if the nutrition of the crop was by ammonium ions, the content of organic acids would be much lower than that with nitrate ions. The nearest crop studied was buckwheat. With this crop the organic acid content with ammonium nutrition was one-fifth of that when given its nitrogen as nitrate.

Dr. DeKock stated that rhubarb and spinach had been studied on the continent of Europe and it had been found that supplying nitrate nitrogen resulted in high oxalate contents of these crops.

Dr. J. Tinsley asked about the stage of growth at which the crops described in his paper had been harvested and to what parts of the plants the data referred to. *Mr. Kirkby* replied that the plants were grown for three to four weeks and the data quoted referred to the leaf tissue. The effect of aging of plants had also been studied and it was found that the difference between ammonium and nitrate nutrition also applied to mature plants but that mature leaves were higher in cations and organic acids than young leaves, whatever the source of nitrogen.

Mr. H. T. Davies asked if the authors considered their work had any relation to work on the cation exchange capacities of roots. *Dr. DeKock* replied that the cation exchange capacity of roots was due to free carboxyl groups on the root surface so there was no reason to suppose that there was no relation between the number of these carboxyl groups and the concentration of inorganic acids within the plant but he did not believe that the cation exchange capacity of the roots was the sole arbiter for the chelation and absorption of anions.

Some processes affecting nitrogen in the soil

By J. K. R. GASSER, B.Sc. (Nott.), Ph.D. (Lond.), F.R.I.C.
Rothamsted Experimental Station

MOST of the nitrogen in soil is in organic combination and unavailable to plants. Microbiological mineralization of organic-N provides ammonium-N, which is subsequently nitrified to nitrate-N. These inorganic forms are available to plants and also to those micro-organisms able to use inorganic-N. Both microbiological processes occur simultaneously in soils, so that the rate of formation or of removal of inorganic-N depends on the relative magnitudes of the two processes. The mineralization of organic nitrogen in soil was extensively reviewed by Harmsen and Van Schreven[59], and Winsor[122] discussed mineralization and immoblization. The results of a detailed investigation using N^{15} to follow mineralization-immobilization relationships were described by Jansson [62]. As both mineralization and immobilization are microbiological processes they occur most readily under conditions favouring activity of the micro-organisms, for example in moist, warm, well-aerated soils. The optimum conditions are moisture content of 50–60 water-holding capacity, temperatures of 30–35 °C, and soil aggregates not exceeding about 3.5 mm diameter. Organic matter accumulates under acid conditions. On liming acid soils the amount of extra mineral-N formed on incubation increases with decreasing acidity of the soil to about pH 6·5[122]; about this value no extra N is mineralized by liming. Micro-organisms, both those mineralizing N and those immobilizing N, require an available source of carbon for energy, but its C/N ratio is critical. For example, adding carbohydrate immobilizes nitrogen, whereas adding plant material with a C/N ratio less than about 30:1 does not immobilize nitrogen, and as the ratio decreases nitrogen becomes increasingly mineralized.

Fixation of ammonium and ammonia in soils

Ammonium fixation

When ammonium salts are added to soils, the ammonium ion is normally held on the base-exchange complex, and may be removed by leaching with a salt solution (e.g., barium chloride). In some soils, usually those containing illite, ammonium (or potassium) may enter the clay lattice and no longer be released on leaching with a salt solution. The ion is then described as 'fixed'. Jansson[62] reviewed the work on ammonium fixation:

That chemical ammonium fixation by mineral soil may be an interfering factor in biological nitrogen transformations has only recently been considered. From some observations that ammonium nitrogen fixed by synthetic zeolites could not be utilized by barley and oat plants, Pfeiffer *et al.*[88,89] concluded that zeolitic soil minerals may be responsible for a fixation process rendering ammonium nitrogen unavailable to plants. Their suggestions were not followed up, however, and later observations on chemical ammonium fixation[72,18,85] were merely concerned with the chemical extraction and determination of ammonium nitrogen in soil. About twenty years ago the occurrence of potassium fixation in soils was demonstrated[109] and it was soon revealed that there is an important relationship between potassium and ammonium fixation[99,16]; McBeth[72] had already noticed that treating the soil with a potassium salt decreased ammonium fixation.

As to the biological importance of the chemical ammonium fixation, Chaminade and Drouineau[36], and Chaminade[35] concluded that the fixed ammonium nitrogen is not available to plants.

During recent years chemical ammonium fixation has been studies mainly in the U.S.A.[4,5,6,7,8,19,20,56,57] and in Japan[11,12,13,58] but also by workers in some other countries[78,79,81,92]. From these investigations some important general features of the fixation process have been revealed, which may be summarized as follows:

The parallelism between potassium and ammonium fixation is substantiated. Potassium and ammonium are fixed by the same mechanism in the soil; ammonium fixation prevents subsequent potassium fixation and vice versa. According to Nommik[81] potassium seems even to hold a key position in determining the extent and firmness of this kind, while others show practically none. The fixation process is a function of the mineral colloid of the soil.

The fixation process involves a cation exchange reaction and will diminish the cation-exchange capacity of the soil. The fixed ions will occupy positions, which normally contribute to the exchange capacity[90].

Subsoils normally have higher fixation capacity than the corresponding topsoils.

Drying the soil after addition of ammonium or potassium will increase fixation; repeated wetting and drying will increase fixation still more.

Liming or otherwise decreasing soil acidity will normally increase fixation.

The biological availability of fixed ammonium nitrogen is low.

The mechanism of ammonium fixation

As to the mechanism of the fixation process two main theories have been postulated; both theories contain concepts of importance in understanding the experimental findings.

According to one view, the fixed ions are supposed to enter into the lattices of the fixing minerals with the formation of new, somewhat modified minerals or even the restoration of micaceous minerals originally present in the soil but subsequently decomposed under the influence of weathering and the action of plants[90]. The last-mentioned possibility, of course, refers only to potassium fixation.

The second view is that the fixation process is in principle the same as the normal cation exchange process, the only difference being that a proportion of the potassium and ammonium ions enter into exchange positions from which they are not displaced by other cations in the normal way of cation exchange.

The fixation has been related to the size of the unhydrated ions[86]. If the size of the ions is close to that of certain free spaces within the mineral lattices, the ions will become trapped in these spaces. Alternatively, according to Truog and Jones[103], the fixed ions are trapped in interlayer exchange positions of expanding lattice minerals. In this way fixation will be enhanced if the lattices are caused to contract during the fixation process or soon afterwards. Allison and Roller[8] attribute fixation mainly to the minerals vermiculite, illite, montmorillonite and biotite, and stress the contracting action of potassium on the mineral lattices. Hendricks and Dyal[60] also found that potassium saturation contracted the lattice of montmorillonite, and nontronite.

Peech[87] claims that ammonium ions exert a pronounced blocking effect on the conversion of non-exchangeable potassium into the exchangeable form. It is also interesting to note that Allison and Roller[8] found a higher availability to plants of the fixed ammonium after the exchangeable ammonium was removed by leaching with calcium than by leaching with potassium. Calcium was supposed to expand the mineral lattices, potassium to contract them. Furthermore, Aomine and Wada[13] in studying the ammonium fixation capacity of homoionic soil found decreasing fixation capacity when the soil was saturated with the following ions: Na > Ca > Li > Mg > Ba > H > K. This order may be related to the ability of the different ions to expand the lattices and make the interlayer adsorption sites available for the ammonium ions. For ammonium fixation similar results are reported by Harada and Kutsuna[58] and for potassium fixation by Wiklander[121].

Reactions between ammonia and clay or organic matter in soils

(a) *Clay*

The NH_3 sorbed and/or retained by dry bentonite, halloysite and synthetic gibbsite at pressures from 0–100 mm of mercury were investigated by Brown and Bartholomew[27] and they found that in the range pH 8 to 10 the amount of NH_3 sorbed and chemisorbed by Ca bentonite did not differ. Synthetic gibbsite, Al $(OH)_3$, did not chemisorb NH_3 presumably because of the basic nature of this mineral. They suggested that chemisorbed NH_3 is:

(i) that which has gained a proton and undergone exchange reactions with the exchangeable cations and

(ii) that which has reacted with the weakly dissociated hydroxyl groups of the lattice edges. In further work, Brown and Bartholomew[28], investigated the ammonia sorption by homoionic clays containing small amounts of adsorbed water vapour. At ammonia pressures below about 80 mm Hg, 'dry' bentonite and halloysite sorbed more ammonia than comparable moist clays. At higher ammonia pressures, moist clays sorbed more ammonia than 'dry' clays with most sorption at higher moisture levels. Aqueous vapour apparently competed with ammonia for sorption sites on the clays. The kind of cation on the exchange complex also exerted an influence on ammonia sorption by moist clay, probably by altering the quantity and nature of the water of hydration. Exchangeable cations influenced ammonia sorption in the same order with both moist and dry clays, i.e., Al-clay > Ca-clay > K-clay.

Chao and Kroontje[37] found that at low concentrations ammonia adsorption on all soils followed Langmuir's monomolecular adsorption theory. The differential slopes obtained for different sections of the curves were related to reactions involving 'hydrogen' exchange sites and physical adsorption. The S-shaped curve obtained for high concentrations suggests the formation of poly-molecular layers. The ammonia adsorption on neutral and basic soils, however, also gave typical Freundlich adsorption isotherms and furthermore closely simulated the behaviour of a buffer medium.

(b) *Organic matter*

Burge and Broadbent[29] found that ammonia fixation in organic soils was linearly correlated with the carbon content. In the presence of oxygen 1 molecule of ammonia was fixed per 29 atoms of carbon, and 1 molecule of ammonia for every 45 atoms of carbon in the absence of oxygen. Treating these organic soils with ammonia usually decreased their capacity to retain barium and cupric ions, but the amount of ammonia fixed was not related to the decrease in cation retention. Blocking hydroxyl groups in soil organic matter by treating with dimethyl sulphate decreased the capacity of the organic matter to fix ammonia, indicating that these groups are involved in the fixation reaction. Aldehyde groupings are apparently not involved. Formation of amides through the action of anhydrous ammonia on carboxylic esters may play a minor role in ammonia fixation. A tracer experiment showed that fixed ammonia is slowly made available to plants.

Broadbent *et al*[24] found that ammonium salts and ammonium hydroxide added to soils in low concentrations react with organic matter to fix the nitrogen in a non-exchangeable form, both in the presence and in the absence of oxygen. They and Shoji and Matsumi[96] found little ammonia was fixed in acid soils and fixation increased with pH in alkaline soils. Some pure polyphenolic compounds fix ammonia in the presence of oxygen, to form a stable organic matter nitrogen complex[24]. Very little of a lignin-ammonia complex was mineralized on incubation[96].

Nommik and Nilsson[82] treated various mineral and organic soils with ammonium hydroxide, calcium cyanamide, or urea. After one to three days the N not removed by leaching was considered as fixed. Fixation of NH_3 increased with increasing oxygen consumption, and with increasing pH. At pH > 10, net fixation of NH_3 was affected by simultaneous non-enzymatic de-amination of organic-N in the soil and by dissolution of organic matter. Fixation of N from calcium cyanamide was more than double that from ammonium hydroxide. The ammonium fixing capacity increased with increasing organic carbon content, organic soils fixed more than mineral soils, and low humified virgin peats twice as much as cultivated mucks. Whereas Shoji and Matsumi[96] found that in peat soils the amount of NH_3 fixed was not related to the kind and degree of humification and that fixation decreased in the presence of cations in the order $K^+ > Ca^{++} > Al^{+++}$.

Leaching and capillary rise of nitrate

Various workers[39,44,45,46,95] have studied the movement of mineral nitrogen from the surface of soils and found that in light sandy soils, where vertical structural fissures are absent, and in a freely-drained clay-loam, nitrate is distributed down the profile on leaching. In the heavy soils which are well fissured and possess a gleyed horizon of partially decomposed plant debris, nitrogen may be lost by leaching or denitrification after removal from the cultivated layer. All workers found that the amount of rain required to remove nitrate from the surface layers differed little between light and heavy soils, and that heavy and continuous rain is required to remove nitrate completely from either soil. Ammonium nitrogen is not leached from the surface layer of either light or heavy soil.

In Queensland, Wetselaar[119] applied chloride and nitrate to the surface of a clay soil and followed their movement into the sub-soil during wet weather. He found that they moved downwards 1·1 in. for each inch of rainfall. Movement out of the topsoil was enhanced by channels left by partly decomposed roots. Sodium chloride applied to sandy soil moved more for each unit of rain than in clay soil, except that a sandy soil with much silt in the surface layer behaved more like the clay presumably because of a low infiltration rate.

In France[43], in Kenya[91], in Uganda[100], and in Queensland[117] nitrate accumulated in the surface 2 in. of soil in dry weather due to upward movement particularly from the 2–6 in. layer. When the soil was wetted once during the dry season nitrate accumulated in the 0–1 in. and 1–2 in. zones and decreased in the 6–9 in. and 9–12 in. layers during the subsequent drying.

Wetselaar[118] applied chloride and nitrate to leach during the wet season and followed their movement during the following dry season. Generally, during drying phases nitrate and chloride increased in the surface layers and decreased in the sub-surface layers; rain reversed this. Both anions behaved similarly, and nitrate tended to increase immediately after rain due to biological formation in the rewetted zone. During drying, most of both anions accumulated in the 0–1 in. layers on two soil types and in the 1–2 in. layer on the third; in all soils, this accumulation was below the bottom of the surface crust where the physical continuity of the soil was broken.

When rainfall is insufficient to cause complete leaching, nitrate may accumulate at depth[64,119]. Although some nitrate may move upwards in the surface layers during drying of the soil, Stephens[100] did not consider that nitrate accumulated at depth could move to the surface.

Gaseous losses of nitrogen

By volatilization of ammonia

Ammonia is lost from ammonium salts applied to alkaline soils. Urea is hydrolysed in soil to form ammonium carbonate increasing the pH; when it becomes greater than pH 7 ammonia may be lost, so that losses may occur from urea applied to acid as well as to calcareous or alkali soils. Gasser[48] recently reviewed the use of urea as a fertilizer, including factors affecting losses of ammonia from urea applied to soils and grass swards: Hydrolysis of urea increases the pH of soils, and above pH 7 ammonia may volatilize. The most important factor affecting the loss of ammonia from soils is considered to be their base-exchange capacity (B.E.C.)[10,47]. Considerable losses may be expected from soils whose B.E.C. is less than 10 meq/100 g; the probable maximum loss decreased from 20 per cent at 10 meq/100 g to 10 per cent at 20 meq/100 g when 100 lb N/acre is applied as urea to the surface of moist soils. The amounts of ammonia lost also depend on other factors. Increasing pH of soils increases losses, because more ammonia is lost from limed than from unlimed soils[10,42,75,108]. More ammonia is lost when soils are dried[108,113] than when they are maintained at constant moisture content. Maintaining soils at 25 °C, compared with 5–10°, (i) increases losses of ammonia from urea applied to some soils, but decreases losses from others and (ii) decreases the period of loss from 30–70 days to 10–14 days. The physical form of the urea does not affect ammonia losses from surface-applied material because similar amounts are lost from soils with pelleted urea, crystalline urea, or urea in solution.

Losses of ammonia may be decreased either by mixing urea with the soil[42,75,84] or by mixing urea with acid materials such as superphosphate. Tripathi[102] found that adding organic materials to soil also decreased losses of ammonia from urea applied to the soil. When applied to the surface of calcareous soils, more ammonia is lost from ammonium sulphate than from urea[47,71,107], although the limited evidence available[47] suggests that on mixing the fertilizers with soil more ammonia is lost from urea than from ammonium sulphate.

Considerable amounts of ammonia may be lost from urea applied to turf. The greatest percentage loss recorded is 36 per cent of 100 lb N/acre applied to Bermuda grass[108] other losses vary from 2–30 per cent of N applied at rates from 50–300 lb N/acre[61,66,97,107].

More ammonia is lost when urea is applied to soil covered with plant debris than when applied to soil alone[10,74].

Although very little or no ammonia is lost from ammonium salts applied to slightly acid soils, losses from ammonium sulphate applied to alkaline soils increases markedly with pH[63]. The anion also affects the amounts lost, because Nommik and Warlin[83] and Vanstallen[104] found that more ammonia was lost from ammonium sulphate than from ammonium nitrate supplying equivalent quantities of ammonium; Larsen and Gunary[67] found that twice as much ammonia was volatilized from ammonium sulphate as from mono- and di-ammonium phosphates and ammonium nitrate and negligible amounts from magnesium ammonium phosphate, applied to calcareous soils.

By losses as nitrogen and oxides of nitrogen

The process of denitrification leading to losses of nitrate as gaseous products has long been known, and was discussed by Bremner and Shaw[21]:

The term denitrification is now reserved to describe the microbial process whereby nitrate is reduced to gaseous nitrogen compounds such as nitrous oxide and nitrogen. It has been used to describe other processes involving the disappearance of nitrate such as the assimilatory reduction of nitrate accomplished by higher plants and micro-organisms to meet their requirements for organic nitrogen, but denitrification differs from these in that it leads to gaseous loss of nitrogen.

All denitrifying bacteria so far studied have been found to be facultative anaerobes. Provided an oxidizable substrate is present in the medium, they can grow under anaerobic conditions in the presence of nitrate or under aerobic conditions in the presence of any suitable source of nitrogen. The oxidizable substrate can be inorganic as well as organic, e.g. autotrophic bacteria have been described which can grow anaerobically in the presence of nitrate at the expense of molecular hydrogen[77,106] or of sulphur, sulphide or thiosulphate[76,17,15]. These observations are concordant with the theory, first proposed by Weissenberg[114,115], that microbial denitrification is equivalent to aerobic oxidation with nitrate replacing oxygen as the ultimate oxidising agent, and it is now accepted that, in the process of denitrification, nitrate instead of oxygen serves as the final hydrogen (or electron) acceptor in enzymic dehydrogenations of the organic or inorganic substrate. The mechanism of denitrification has not been fully elucidated.

The demonstration by Gayon and Dupetit[50] that soil contains bacteria capable of reducing nitrate to molecular nitrogen and nitrous oxide led to much speculation concerning the loss of nitrogen from soil by denitrification and considerable alarm was caused in agricultural circles when Wagner[110] published the results of an extensive investigation from which he concluded that serious loss of nitrogen may occur when nitrate is used as a fertilizer. This conclusion was soon challenged by others who pointed out that it is unusual to add organic matter and nitrate to soil at the same time as was done in Wagner's experiments, and it subsequently became generally accepted that loss of nitrogen by denitrification is negligible in well-aerated soils provided no addition of organic matter is made when nitrate is applied. Further work tended to confirm this view, and when Waksman[(111)] reviewed work on denitrification his conclusion was that 'the phenomenon of denitrification is of no economic significance in well-aerated, not too moist soils, in the presence of moderate amounts of organic matter or nitrate'. This conclusion was not seriously questioned until Broadbent[(23)] published the results of an investigation on denitrification in Californian soils which indicated that denitrification is of considerable economic significance in aerated soils in the presence of moderate amounts of organic matter and nitrate. For example, he found that when a sandy loam containing 1290 ppm of total nitrogen and 64 ppm of nitrate-N was incubated in an aeration stream containing 21 per cent oxygen it lost 250 ppm of nitrogen in seven days. Subsequent work by[(69,25,105,120,14,112)] has lent support to the view that significant loss of nitrogen from soil may occur by the process of denitrification. Further support can be taken from soil nitrogen balance sheets obtained from lysimeter experiments[(1)] as they frequently show substantial losses of nitrogen that are difficult to explain unless it is assumed that considerable loss of soil nitrogen occurs by gaseous evolution.

The study of denitrification in the field is beset with experimental difficulties that have so far proved insuperable and most of the information now available has been obtained from laboratory experiments in which denitrification has been studied either by estimating the gaseous products of denitrification or by following loss of nitrogen by total nitrogen analysis. To obtain accurate results by either of these methods without using labelled nitrate, it is necessary to add nitrate to soil in amounts considerably greater than those usually applied in the field. A further disadvantage of these methods when unlabelled nitrate is used is that if denitrification is accompanied by nitrogen fixation they will measure only the net loss of nitrogen resulting from these two processes. However, it appears unlikely that significant fixation of nitrogen will occur during the process of denitrification since various workers[(41)] have reported that nitrogen fixation is suppressed by small concentrations of nitrate.

Although the early work established the phenomenon of denitrification, many ideas have been developed only since 1951. Some of the results appear to differ in the importance of various factors affecting denitrification. The following summaries of seemingly important work on this subject do not attempt to assess critically the information on individual points but present the main findings of investigations grouped under their subject matter.

(a) *The mechanism of denitrification*

The mechanism of the reactions leading to the gaseous products of denitrification, nitric oxide, nitrous oxide and nitrogen have been investigated and also the effects of various factors on the proportion and amount of each. In solutions containing ammonium and nitrite ions, Smith and Clark[(98)], found that the reaction between these ions to yield gaseous nitrogen was promoted by acid conditions and a nitric oxide atmosphere. In the presence of oxygen, nitrite tended to be converted to nitrate much more than to react with ammonium, and in soils, oxidation was even more rapid. In soils, the reaction between the ammonium and nitrite ions to form nitrogen gas did not appear to be nearly as important as the reduction of nitrite to nitrogen gas by some component of the soil complex. Evolution of nitrogen and nitrous oxide and diminution of nitrite content without concomitant diminution of substrate ammonium were observed. The presence of soil interfered with the ammonium/nitrite reaction. No nitrogen was lost as nitric oxide or nitrogen di-oxide from moist and aerobic soil. Wijler and Delwiche[120] identified nitrous oxide, nitrogen and nitric oxide as products of denitrification, and found that nitrous oxide could be reduced to nitrogen under suitable conditions. Nitric oxide was also further reduced, and nitrous oxide was the major product of denitrification under most soil conditions. Greater moisture contents of the soil indirectly increased denitrification by decreasing the diffusion of oxygen into the soil. Below the permanent wilting point, moisture content directly decreased denitrification by limiting microbial activity. Total denitrification rates were quite constant above pH 6, but the proportion of nitrous oxide and nitrogen depended on pH. Above pH 7, nitrous oxide was readily reduced to nitrogen, between pH 6 and 7 reduction was inhibited. Below pH 6, nitric oxide was formed and the rate of denitrification decreased. Oxygen suppressed denitrification; tensions of 5 mm Hg decreased denitrification to about one tenth of the anaerobic rate. The initial nitrate concentration did not affect the rate of denitrification. Using straw instead of lucerne as an energy source decreased the rate of denitrification, but the proportions of nitrogen and nitrous oxide formed were unchanged. Using the stable isotope N^{15} to label the added nitrate, denitrification products were shown to be wholly from the nitrate, so that a van Slyke reaction of nitrous acid did not occur. Little nitrate was reduced to ammonia. Rapid oxidation of ammonia to nitrate may result in more nitrogen loss under fluctuating oxygen tensions than under consistently anaerobic conditions.

In a study of the sequence of gaseous denitrification products Cady and Bartholomew[30] found that the product to appear initially was nitric oxide, which did not exceed 5 per cent of the labelled nitrogen. The presence of nitric oxide in the gaseous phase was closely associated with the presence of nitrite in the soil and was presumed to have arisen from the decomposition of nitrous acid. After the nitric oxide, nitrous oxide appeared in the gaseous phase, with increasing nitrous oxide, nitric oxide decreased and eventually disappeared. Following the increase in nitrous oxide, nitrogen appeared and eventually accounted for about 90 per cent of the labelled nitrate with a concomitant decrease and disappearance of nitrous oxide. On investigating the equilibrium reaction between nitric oxide and non-gaseous compounds in the denitrification process in soil they found that nitric oxide rapidly equilibrates isotopically with non-gaseous products during the early stages of denitrification[32]. They also found that under acid soil conditions nitric oxide appeared in the gaseous phase, whereas under basic soil conditions nitric oxide did not appear from denitrification reactions although added nitric oxide persisted in the system and isotopically equilibrated with the non-gaseous compounds in the soil. The equilibrium reactions between nitric oxide and the non-gaseous compounds in the denitrification system can not be adequately explained by the conventional equation

$$3NO_2^- + 2H^+ \rightleftarrows NO_3^- + 2NO + H_2O$$

and other reactions and products may be involved.

(b) *Factors affecting anaerobic denitrification in soils*

Bremner and Shaw[21] found that under waterlogged conditions unamended soils did not all denitrify but that rapid denitrification in soil could be induced by adding organic materials such as glucose. They considered that denitrification can be followed by total-N analysis if the organic material used to induce denitrification is not added in such excess that it also promotes significant fixation of atmospheric nitrogen. The percentage of added nitrate-N lost by denitrification on incubation of waterlogged soils with different amounts of nitrate and sufficient glucose for denitrification was found to be the same whatever the level of application of nitrate, and that it was accompanied by a rapid but temporary accumulation of large quantities of nitrite and by the formation of small amounts of ammonia. Hydroxylamine could not be detected during denitrification, but this compound was rapidly decomposed in the soils examined by an apparently purely chemical process. As denitrification of nitrate in soil is a microbiological process, it depends on factors such as pH, temperature water content of the soil, and Bremner and Shaw[22] found that denitrification was slow below pH 4·8 increasing with pH to be rapid at pH 8·0; the rate of denitrification increased rapidly with temperature from 2°C to 25°C, with a maximum rate at about 60°C; below 60 per cent W.H.C., there was no denitrification, between 60 per cent and 70 per cent W.H.C., there was some denitrification and above 70 per cent W.H.C., denitrification increased with increasing moisture content. There were no losses with continuous aeration. The rate of denitrification depends on the amount and type of organic material present, and under conditions conducive to denitrification about 80 per cent of nitrate-N added to Rothamsted soils was lost by denitrification in five days.

Nommik[80] similarly found that:

(i) at moisture contents less than 60–70 per cent W.H.C. denitrification depended on the partial pressure of oxygen,

(ii) denitrification was rapid in the presence of easily oxidizable organic material,

(iii) nitrate concentration affected little the rate of denitrification, at higher concentrations nitrous oxide was the major reduction product,

(iv) denitrification was negligible at 3°C, and increased to a maximum at 60–70°C,

(v) denitrification was more rapid in neutral soils than those below pH 6, and under acid conditions nitrous oxide was not reduced to nitrogen,

(vi) very little ammonia was formed during denitrification.

Nommik also studied the composition of the gasses evolved during denitrification and found that nitric oxide was formed at low pH in the presence of much nitrite, and also by reduction of nitrite in the presence of ferrous ions. Using material labelled with N^{15}, molecular nitrogen formed in the reduction of nitrate was shown to have been nitrous oxide before the final reduction, and that direct reduction from nitrite to nitrogen did not occur.

In some Australian soils, McGarity[73] also found that the rate of denitrification increased with increasing moisture content of the soils. Added glucose markedly increased denitrification in soil containing little organic carbon but much less in those with more native organic matter. Increasing nitrate-N from

60 to 472 ppm did not affect either the rate or pattern of denitrification, which was not prevented by the absence of micro-organisms in any of the soils used. The denitrifying activity of red-brown earths from rotation plots was greatest in the order permanent pasture, continuous wheat, fallow-wheat-pasture-pasture, depending on the total soil carbon and its availability as an energy substrate.

In Ghana, Greenland[54] found that only small and insignificant losses of nitrogen occurred in unsaturated soil (70 per cent W.H.C.) to which nitrate with or without glucose had been added; larger and rapid losses occurred from saturated soils. Soils from under standing forest or old grassland denitrified without adding a carbon energy source, those cultivated or from young grassland needed added glucose for denitrification.

Using infra-red analysis to determine nitrous oxide contents of soil air collected both in laboratory experiments and in the field, Arnold[14] found that soils approaching saturation with moisture rapidly released much nitrous oxide; at lower moisture contents it was occasionally evolved but much more slowly.

Jones[65] studied the losses of nitrogen from moist soils maintained anaerobic by reduced pressure, and found that more than 80 per cent of nitrogen added as nitrate was lost in 3 days; losses were similar with and without an added energy source. Using N^{15} labelled material, over the range pH 4·0–6·0 only a negligible trace of the evolved nitrogen came from sources other than nitrate or nitrite formed in or added to the soil.

Schwartzbeck *et al*[93] studied the nitrogen gasses evolved from four soils and found that more nitrogen and nitrous-oxide were evolved from clay than from silt and sandy soils. More nitrous oxide was formed from ammonium nitrate and more nitrogen from nitric acid. Little nitrogen or nitrous oxide was evolved from soils treated with ammonium chloride or urea. The percentage loss was the same at high and low rates, and losses were greater when the soils were saturated than at field capacity.

(c) *Aerobic denitrification in soils*

In some Californian soils Broadbent[23] found that nitrogen was lost at all oxygen levels when both nitrate and ground clover were added. Below 4·5 per cent O_2 all nitrate disappeared from untreated soils. Maximum denitrification was at pH 6·3 with less in more acid and in more alkaline soil. Evidence was obtained that oxygen in nitrate was used to oxidize organic matter when the aeration stream contained less than 0·14 per cent O_2. Part of the large losses employed in this study was probably due to aerobic denitrification. Following this work, Broadbent and Stojanovic[25] found that negligible amounts of nitrate were reduced to ammonium at all oxygen concentrations though slightly more occurred under anaerobic conditions than aerobic. Denitrification of added nitrate was inversely related to partial pressure of oxygen but was appreciable even under fully aerobic conditions. No nitrogen was lost from added ammonium salts.

Allison *et al*[3] found that with soils at a moisture content of one-third atmosphere only traces of nitrogen were lost to the air either with or without carbon sources. On aerating soils with an atmosphere containing 2·3 per cent oxygen about half the soils lost nitrogen; with 0·5 per oxygen, 10 per cent of the nitrate was lost from soils alone and 50 per cent on adding 0·5 per cent glucose. They suggest that although appreciable denitrification did not occur until the oxygen partial pressure was much reduced, under field conditions some denitrification would be expected at times even in open, well-drained soils. In further investigations on denitrification in well aerated soils Carter and Allison[33] found that very little if any nitrogen was lost under aerobic conditions except with added dextrose, when losses were 0–12 per cent. They concluded that denitrification is of minor importance in soils that are kept strictly aerobic, but under field conditions this ideal condition may not exist at all times even in the more open soils. Cady and Bartholomew[31] incubated soils at constant partial pressures of oxygen to study its effect on 'aerobic denitrification'. Decreasing oxygen pressure from about 8 per cent to 5 per cent to 1 per cent each increased denitrification five fold, so that it was twenty-five times as great at 1 per cent as at 8 per cent, showing that gaseous oxygen was used before oxygen combined as nitrate. The amount of denitrification was closely associated with the level of the biological pressure or need of oxygen by the microbial population and nitrate would only be appreciably reduced to gaseous forms when there was adequate available carbon, and less than 7 per cent oxygen by volume. Nitrous oxide was produced in the aerobic system.

(d) *Losses during nitrification of ammonium sulphate*

Gerretsen and De Hoop[51] found that much nitrogen was lost from ammonium sulphate during nitrification in an acid liquid culture or in sandy soils when the acidity dropped below pH 5·5. No nitrogen was lost on adding nitrate or when the soil remained above pH 5·5. Similarly on adding ammonium sulphate to a sandy loam at concentrations comparable to those expected in fertilizer bands, Carter and Allison[34] found that 0–30 per cent of the added nitrogen was lost in forms other than ammonia. In these studies of limed and unlimed soils where the pH varied from 4·7 to 6·9, the losses increased with increase in added ammonium sulphate and were greater on limed than on unlimed soil.

Recent work by Currie[40] on soil aeration and Greenwood[55] on decomposition of organic materials in water-saturated crumbs provide explanations of dentrification in apparently aerobic soils by the existence of anaerobic zones. Currie points out that, as soil crumbs approach saturation, diffusion is confined to the liquid phase where the diffusion co-efficient is four orders of magnitude lower than in the gaseous phase. Therefore, within larger crumbs critical values of oxygen and carbon di-oxide are reached and their centres become effectively anaerobic. From perfusion experiments, Greenwood considered that the changeover from aerobic to anaerobic metabolism occurs at oxygen concentrations less than 3×10^{-6} M; for a loam soil the critical diameter is about 0·35 cm. Currie discusses the effects of anaerobic centres in the crumbs:

Lack of oxygen upsets the balance of nutrient uptake and decreases the permeability of the roots to water. Root elongation is inhibited, and so the search for food and water stops. This effect is greatest in soils where mechanical impedance to the movement of the root is high. Indirect effects are the changes in nutrient status, for example, the change of iron and manganese to the ferrous and manganous forms and the reduction of nitrates and sulphates to useless and even toxic forms.

High carbon dioxide concentration may accentuate these effects or act independently. It may be directly toxic to the root cells, or indirectly by the low pH it produces within the microstructure. For each tenfold increase in the CO_2 concentration in aqueous solution there is a fall of about half a pH unit. In this way, a soil which shows satisfactory pH values on normal analysis may be prone to unsuspected pH troubles, particularly if its buffer capacity is low. By-products of anaerobic respiration may be toxic to plants and, under long anaerobic periods, the natural compounds which stabilize soil structure may be degraded to forms which no longer fulfil this function. Essentially, the anaerobic zone has the effect of decreasing still further the availability of nutrients already decreased chemically. If, for example a crumb of a certain radius is just wholly aerobic, a crumb twice that size will have 30 per cent of its volume anaerobic, that is, 'availability' is decreased by 30 per cent. A crumb of ten times the critical radius would be 84 per cent anaerobic.

Aeration is thus relevant in deciding the fineness of soil tilth. Obviously the tilth must not be so fine that at field capacity the soil remains waterlogged, or both trafficability and diffusion suffer. At the other limit, the needs of good crumb aeration rather than ease of cultivations may determine the maximum crumb size.

The use of a nitrification inhibitor to decrease losses of nitrogen by leaching and denitrification

In soil, nitrite and nitrate nitrogen are susceptible to loss by leaching and denitrification; ammonium nitrogen moves little in soil and does not react to yield gaseous products which are lost from the soil, except that ammonia may be lost when the soil is alkaline. The risks attached to nitrate nitrogen appear to be greater than those to ammonium nitrogen as a nutrient source for plants, therefore maintaining applied nitrogen in the ammonium form should improve its efficiency compared with nitrate under conditions where the latter may be lost by leaching and/or denitrification. A number of chemicals such as thiourea, and dicyandiamide, as well as those chemicals used as soil sterilants, inhibit nitrification, affect other micro-biological processes, and are phytotoxic to plants. Recently, a compound, 2-chloro-6-(trichloromethyl)-pyridine inhibiting the biological transformation of ammonium to nitrite was developed by the Dow Chemical Co. under the name 'N-Serve'[52,53]. This inhibitor has little or no effect on other soil microbiological processes[94].

Ammonium from fertilizers treated with this compound would not be nitrified and the risks of losses by denitrification or leaching are removed so that fertilizer efficiency may be increased. In the U.S.A., Swezey and Turner[101] compared ammonium compounds and urea with and without the inhibitor as fertilizers for irrigated cotton, sweet corn and sugar beets when crops yielded more with treated than with untreated fertilizers.

In field experiments in Britain[49], ammonium sulphate alone and treated with the nitrification inhibitor 2-chloro-6-(trichloromethyl)-pyridine, at rates equivalent to 1 per cent and 2 per cent of the weight of N applied, was applied to sandy loam and clay loam soils. Ammonium sulphate was given in November 1962, and soil samples taken in spring 1963, to a depth of 36 in. on the sandy loam and 18 in. on the clay loam, were analysed. When the ammonium sulphate was broadcast on the soil surface, 2 per cent of the inhibitor, inhibited nitrification of the ammonium-N more than 1 per cent, but when it was placed in the soil both rates were equally effective.

Untreated ammonium sulphate was applied in spring to previously unfertilized plots, and ryegrass was sown on all plots. On the sandy loam soil, yield and N uptake were increased most by the N applied in spring; of the autumn-applied dressings, treated fertilizer placed in the soil increased yield and uptake most and untreated fertilizer broadcast on the surface least. On the clay loam soil, spring-applied fertilizer, and autumn-applied treated fertilizer placed in the soil increased yields of dry matter most, but ryegrass took up more N from the spring than from the autumn dressings. Untreated fertilizer broadcast on the surface in autumn on this soil anomalously increased yield and N uptake more than treated fertilizer similarly applied, and more than untreated fertilizer placed in the soil.

This material is potentially valuable for investigating the importance of some of the nitrogen transformations in the field, and for comparing the effects of ammonium forms of N with nitrate forms on crop composition and yield.

References

1. ALLISON, F. E. (1955). 'Advances in Agronomy', Vol. VII, p. 213. Academic Press Inc., New York.
2. ALLISON, F. E. (1963). *Soil Sci.*, **96,** 404.
3. ALLISON, F. E., CARTER, J. N. and STERLING, L. D. (1960). *Proc. Soil Sci. Soc. Amer.*, **24,** 283.
4. ALLISON, F. E., DOETSCH, J. H. and ROLLER, E. M. (1951). *Soil Sci.*, **72,** 187.
5. ALLISON, F. E., DOETSCH, J. H. and ROLLER, E. M. (1953). *Soil Sci.*, **75,** 373
6. ALLISON, F. E., KEFAUVER, M. and ROLLER, E. M. (1953). *Proc. Soil Sci. Soc. Amer.*, **17,** 107.
7. ALLISON, F. E. and ROLLER, E. M. (1955). *Soil Sci.*, **80,** 349.
8. ALLISON, F. E. and ROLLER, E. M. (1955). *Soil Sci.*, **80,** 431.
9. ALLISON, F. E., ROLLER, E. M. and DOETSCH, J. H. (1953). *Soil Sci.*, **75,** 173.
10. ANDERSON, J. R. (1962). *Proc. 36th Congr. S. afr. Sug. Tech. Ass.*, 97
11. AOMINE, S. (1951) *J. Sci. Soil, Man. Japan*, **22,** 83.
12. AOMINE, S. and HIGASHI, T. (1953). *J. Sci. Soil, Man. Japan*, (a) **23,** 105; (b) **23,** 185.
13. AOMINE, S. and WADA, K. (1952). *J. Sci. Soil. Man. Japan*, (a) **22,** 227; (b) **22,** 315; (c) **23,** 1.
14. ARNOLD, P. W. (1954). *J. Soil Sci.*, **5,** 116.
15. BAALSRUD, K. and BAALSRUD, K. S. (1954) *Arch. Mikrobiol*, **20,** 34.
16. BARSHAD, I. (1951). *Soil Sci.*, **72,** 361.
17. BEIJERINCK, M. W. (1920) *Proc. Acad. Sci. Amst.*, **22,** 899.
18. BENGTSSON, N. (1924). *Soil Sci.*, **18,** 255.
19. BOWER, C. A. (1950). *Soil Sci.*, **70,** 375.
20. BOWER, C. A. (1951). *Proc. Soil Sci. Soc. Amer.*, **15,** 119.
21. BREMNER, J. M. and SHAW, K. (1958). *J. agric. Sci.*, **51,** 22.
22. BREMNER, J. M. and SHAW, K. (1958). *J. agric. Sci.*, **51,** 40.
23. BROADBENT, F. E. (1951). *Soil Sci.*, **72,** 129.
24. BROADBENT, F. E., BURGE, W. D. and NAKASHIMA, T. (1960). *Trans. 7th int. Congr. Soil Sci.*, **2,** 509.
25. BROADBENT, F. E. and STOJANOVIC, B. F. (1952). *Proc. Soil Sci. Soc. Amer.*, **16,** 359.
26. BROADBENT, F. E. and TYLER, K. B. (1962) *Proc. Soil Sci. Soc. Amer.*, **26,** 459.
27. BROWN, J. M. and BARTHOLMEW, W. V. (1962). *Proc. Soil Sci. Soc. Amer.*, **26,** 258.
28. BROWN, J. M. and BARTHOLOMEW, W. V. (1963). *Proc. Soil Sci. Soc. Amer.*, **27,** 160.
29. BURGE, W. D. and BROADBENT, F. E. (1961). *Proc. Soil Sci. Soc. Amer.*, **25,** 199
30. CADY, F. B. and BARTHOLOMEW, W. V. (1960). *Proc. Soil Sci. Soc. Amer.*, **24,** 477.
31. CADY, F. B. and BARTHOLOMEW, W. V. (1961). *Proc. Soil Sci. Soc. Amer.*, **25,** 362.
32. CADY, F. B. and BARTHOLOMEW, W. V. (1963). *Proc. Soil Sci. Soc. Amer.*, **27,** 546.
33. CARTER, J. N. and ALLISON, F. E. (1960). *Soil Sci.*, **90,** 173.
34. CARTER, J. N. and ALLISON, F. E. (1961). *Proc. Soil Sci. Soc. Amer.*, **25,** 484.
35. CHAMINADE, R. (1940). *C.R. Acad. Sci. Paris*, **210,** 264.
36. CHAMINADE, R. and DROUINEAU, G. (1936). *Ann. Agron.*, **6,** 677.

37. CHAO. T. T. and KROONTJE, W. (1960). *Trans. 7th int. Congr. Soil Sci.*, **2**, 517.
38. CLARK, F. E. and BEARD, W. E. (1960). *Trans. 7th int. Congr. Soil Sci.*, **2**, 501.
39. CUNNINGHAM, R. K. and COOKE, G. W. (1958). *J. Sci. Fd.Agric.*, **9**, 317.
40. CURRIE, J. A. (1962). *J. Sci. Fd.Agric.*, **13**, 380.
41. DELWICHE, C. C. and WIJLER, J. (1956) *Plant and Soil*, **7**, 113.
42. ERNST, J. W. and MASSEY, H. F. (1960). *Proc. Soil Sci. Soc. Amer.*, **24**, 87.
43. GADET, R., SOUBIES, L. and FOURCASSIE, F. (1961). *C. R. Acad. Agric. Fr.*, **47**, 897 (*Soils and Fert.* **25**, 1298).
44. GASSER, J. K. R. (1959). *J. Sci. Fd.Agric.*, **10**, 192.
45. GASSER, J. K. R. (1961). *J. Sci. Fd.Agric.*, **12**, 375.
46. GASSER, J. K. R. (1962). *J. Sci. Fd.Agric.*, **13**, 367
47. GASSER, J. K. R. (1962). Ph.D. Thesis, Univ. of London.
48. GASSER, J. K. R. (1962) *Soils and Fert.*, **27**, 175.
49. GASSER, J. K. R. (1965) *J. agric. Sci.*, **64**, 229.
50. GAYON, U. and DUPETIT, G. (1886). *Mem. Soc. Sci. phys. nat. Bordeaux.* ser. 3 **2**, 201 (quoted by Bremner and Shaw, 1958).
51. GERRETSEN, F. C. and DE HOOP, H. (1957) *Can. J. Microbiol.*, **3**, 359.
52. GORING, C. A. I. (1962). *Soil., Sci.*, **93**, 211.
53. GORING, C. A. I. (1962). *Soil Sci.*, **93**, 431.
54. GREENLAND, D. J. (1962). *J. agric. Sci.*, **58**, 227.
55. GREENWOOD, D. J. (1961). *Plant and Soil*, **14**, 360.
56. HANWAY, J. J. and SCOTT, A. D. (1956). *Soil Sci.*, **82**, 379.
57. HANWAY, J. J., SCOTT, A. D. and STANFORD, G. (1957) *Proc. Soil Sci. Soc. Amer.*, **21**, 29.
58. HARADA, T. and KUTSUNA, K. (1954). *Bull. Nat. Inst. Agric. Sci.Japan* ser. B. No. **3**, 17.
59. HARMSEN, G. W. and VAN SCHREVEN, D. A. (1955) 'Advances in Agronomy' Vol. VII p. 300 Academic Press Inc., New York.
60. HENDRICKS, S. B. and DYAL, R. S. (1950) *Trans. 4th int. Congr. Soil Sci.*, **2**, 71.
61. JACKSON, J. E. and BURTON, G. W. (1962) *Agron. J.*, **54**, 47.
62. JANSSON, S. L. (1958) *Kungl. Lantbr. Hogsk. Ann.*, **24**, 101.
63. JEWITT, T. N. (1942). *Soil Sci.*, **54**, 401.
64. JEWITT, T. N. (1956). *J. Agric. Sci.*, **47**, 461.
65. JONES, E. J. (1951). *Soil Sci.*, **71**, 193.
66. KRESGE, C. B. and SATCHELL, D. P. (1960). *Agron. J.*, **52**, 104.
67. LARSEN, S. and GUNARY, D. (1962). *J. Sci. Fd.Agric.*, **13**, 566.
68. MACK, A. R. (1962) *Nature, Lond.*, **193**, 803.
69. MANN, H. H. and BARNES, T. W. (1951). *J. agric. Sci.*, **41**, 309.
70. MARTIN, A. E. (1962) *Commonw. Bur. Pastures and Field Crops Bull.*, **46**, 72.
71. MARTIN, J. P. and CHAPMAN, H. D. (1951). *Soil Sci.*, **71**, 25.
72. MCBETH, J. G. (1917) *J. Agr. Res.*, **9**, 141.
73. MCGARITY, J. W. (1961). *Plant and Soil*, **14**, 1.
74. MEYER, R. D., OLSON, R. A. and RHOADES, H. F. (1961). *Agron. J.*, **53**, 241.
75. MITSUI, S., OZAKI, K. and MORIYAMA, M. (1954). *J. Sci. Soil, Tokyo*, **25**, 17.
76. NATHANSOHN, A. (1902) *Mitt. zool. Sta. Neapel*, **15**, 655.
77. NIKLEWSKI, B. (1914). *Zbl. Bkt.* (Abt. 2), **40**, 430.
78. NOMMIK, H. (1952). *Växtnäringsnytt*, **8**:9, 13.
79. NOMMIK, H. (1954). *Växtnäringsnytt*, **10**:5, 47.
80. NOMMIK, H. (1956). *Acta Agric. Scand.*, **6**, 195.
81. NOMMIK, H. (1957). *Acta Agric. Scand.*, **7**, 395.
82. NOMMIK. H. and NILSSON, K. O. (1963). *Acta. Agric. Scand.*, **13**, 371.
83. NOMMIK, H. and WARLIN, B. (1960). *K. Lantbr Högsk. Ann.*, **26**, 303.
84. OKUDA, A., TAKAHASHI, E. and YOSHIDA, M. (1960) *Soil and Plant Fd.*, **6**, 144.
85. OLSEN, C. (1929). *C. R. Trav. Lab. Carlsberg*, **17**, (15), 1.
86. PAGE, J. B. and BAVER, L. D. (1940). *Proc. Soil Sci. Soc. Amer.*, **4**, 150.
87. PEECH, M. (1948). 'Diagnostic Techniques for Soils and Crops', American Potash Institute, Washington D.C.
88. PFEIFFER, T. and EINECKE, A. (1905). *Mitt. Landw. Inst. Univ. Breslau*, **3**, 299.
89. PFEIFFER, T., HEPNER, A. and FRANK, L. (1908). *Mitt. Landw. Inst. Univ. Breslau.*, **4**, 321.
90. REITEMEIER, R. F. (1951) 'Advances in Agronomy' Vol. III p. 113, Academic Press Inc., New York.
91. ROBINSON, J. B. D. and GACOKA, P. (1962) *J. Soil Sci.*, **13**, 133.

92. RODRIGUES, G. (1954) *J. Soil Sci.*, **5**, 264.
93. SCHWARTZBECK, R. A., MACGREGOR, J. M. and SCHMIDT, E. L. (1961) *Proc. Soil Sci. Soc. Amer.* **25**, 186.
94. SHATTUCK, G. E. JR. and ALEXANDER, M. (1963) *Proc. Soil Sci. Soc. Amer.*, **27**, 600.
95. SHAW, K. (1962) *J. agric. Sci.*, **58**, 145.
96. SHOJI, S. and MATSUMI, S. (1961) *Res. Bull. Hokkaido nat. agric. Exp. Sta.*, **76**, 37 (*Soils and Fert.*, **25**, 40).
97. SIMPSON, D. M. H. and MELSTED, S. W. (1962) *Proc. Soil Sci. Soc. Amer.*, **26**, 186.
98. SMITH, D. H. and CLARK, F. E. (1960) *Soil Sci.*, **90**, 86.
99. STANFORD, G. and PIERRE, W. H. (1947) *Proc. Soil Sci. Soc. Amer.*, **11**, 155.
100. STEPHENS, D. (1962). *J. Soil Sci.*, **13**, 52.
101. SWEZEY, A. W. and TURNER, G. O. (1962). *Agron. J.*, **54**, 532.
102. TRIPATHI, O. N. (1958) *Proc. nat. Acad. Sci. India*, **27A**, 215.
103. TRUOG, E. and JONES, R. J. (1938). *J. Industr. Engin. Chem.*, **30**, 882.
104. VANSTALLEN, R. (1960) *Agricultura, Louvain*, **8**, 91, (*Soils and Fert.*, **23**, 1784).
105. VERHOEVEN, W. (1952). Delft Doct. Thesis.
106. VERHOEVEN, W., KOSTER, A. L. and VAN NIEVELT, M.C.A. (1954). *Leeuwenhook Ned. Tijdschdr.*, **20**, 273.
107. VOLK, G. M. (1959) *Agron. J.*, **51**, 746.
108. VOLK, G. M. (1961). *J. agric. Food Chem.*, **9**, 280.
109. VOLK, N. J. (1934) *Soil Sci.*, **37**, 267.
110. WAGNER, P. (1895). *Dtsch. Iandw. Pr.* No. 92.
111. WAKSMAN, S. A. (1927) 'Principles of Soil Microbiology'. Balliere, Tindall and Cox, London.
112. WALLACE, A. and SMITH, R. L. (1954). *Soil Sci.*, **78**, 231.
113. WATSON, G. A., CHIN, T. T. and WONG, P. W. (1962). *J. Rubb. Res. Inst. Malaya*, **17**, 77.
114. WEISSENBERG, H. (1897). *Arch. Hyg., Berl.*, **30**, 274.
115. WEISSENBERG, H. (1902). *Zbl. Bakt.*, (Abt. 2) **8**, 166.
116. WELCH, L. F. and SCOTT, A. D. (1960). *Soil Sci.*, **90**, 79.
117. WETSELAAR, R. (1961). *Plant and Soil*, **15**, 110.
118. WETSELAAR, R. (1961). *Plant and Soil*, **15**, 121.
119. WETSELAAR, R. (1962). *Plant and Soil*, **16**, 19.
120. WIJLER, J. and DELWICHE, C. C. (1954). *Plant and Soil*, **5**, 155.
121. WIKLANDER, L. (1950) *Soil Sci.*, **69**, 261.
122. WINSOR, G. W. (1958). *J. Sci. Fd.Agric.*, **9**, 792.

Discussion

Dr. G. ap Griffiths asked *Dr. Gasser* to comment on the loss of ammonia from ammonium nitrate/lime fertilizers. *Dr. Gasser* replied that one would expect losses of ammonia if ammonium nitrate and lime were applied to the soil together but that when these materials were granulated the losses appeared to be greatly reduced. Most work on losses of ammonia had been done on pure urea and ammonium sulphate and when the fertilizer was applied in the form of a granulated mixture losses were almost negligible.

Mr. M. R. J. Holmes said that experimental work at Levington supported *Dr. Gasser's* observations on losses of ammonia from ammonium nitrate/lime fertilizers.

Mr. J. Webber asked if the losses of ammonia from urea were reduced by incorporation into compound fertilizer in the same way as losses from ammonium sulphate. *Dr. Gasser* replied that loss of ammonia from urea was reduced by mixing with other materials such as superphosphate and that granulating the mixtures reduced the losses still further.

Nitrogen releasing properties of various types of organic matter

By J. TINSLEY, B.Sc., Ph.D., F.R.I.C., F.R.S.E.
University of Aberdeen

Introduction

Definition of 'Nitrogen Release'

IN the two previous papers it is implied that 'nitrogen release' refers to the transformation of that nitrogen within the complex humus compounds of the soil into simple chemical forms potentially available for plants. Ammonium is generally regarded as the key form in this microbiological process, although it is apparent that simple organic nitrogen compounds such as amino-acids may be absorbed in small quantities by plants, while in fertile cultivated soils with a near neutral reaction it is certain that much of the ammonium produced by deamination of organic compounds is quickly oxidised to nitrate before uptake.

The subject might also be considered in the more strictly chemical sense of the forms of nitrogen released when soils are hydrolysed, or otherwise treated, to degrade their complex organic polymers. A recent bibliography of the Commonwealth Bureau of Soils cites an extensive list of publications on hydrolysable soil nitrogen in the period 1951–63 and it is apparent that some investigators are seeking a correlation between the forms of nitrogen released by purely chemical treatments and by microbiological mineralization. Both these aspects will be considered in the present paper.

The internal nitrogen cycle

The amounts of nitrogen mineralized in a soil cannot be equated with plant uptake under field conditions, or even in pot culture, though the two may be highly correlated. However part of the nitrogen released can be lost to the plant since some may be carried away in solution in the drainage water—principally as nitrate; some may escape in gaseous form as nitrogen or lower nitrogen oxides when conditions favour the microbiological reduction processes of 'denitrification'; some may be resynthesized into fresh microbial tissues resulting in temporary depletion of the soil mineral nitrogen by biological immobilization; while some may be chemically immobilized by capture of ammonium ions within the lattice structures of certain clay minerals. However, the supply of available nitrogen in the soil system is normally augmented by that added from the atmosphere—mostly in the rainfall. These few pounds of nitrogen per acre are of little significance on heavily manured cultivated soils, but they are important on unfertilized upland areas and afforested land.

The mineralization of organic nitrogen and the whole subject of nitrogen transformations in soils was reviewed in great detail by Harmsen and Van Schreven[13] and by Jansson[14]. The latter emphasized that traditional analytical methods of determining inorganic and organic nitrogen in soil samples give only the resultant or net effect of the opposing microbiological processes of mineralization and immobilization and, on the basis of extensive experiments using N^{15} as a tracer he postulated the operation of an internal nitrogen cycle, which, for an open soil system in the field as opposed to a closed system of soil incubated in a flask, can be expressed diagramatically as shown in Fig. 1 and Fig. 2.

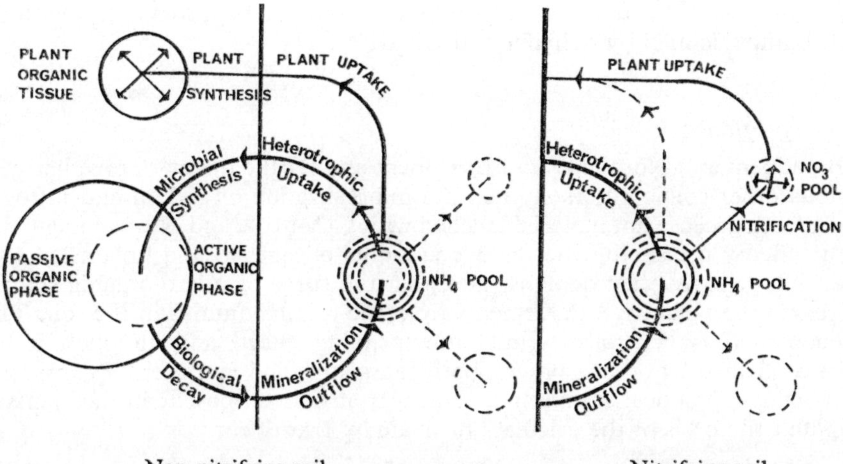

Fig. 1. Relations between the internal nitrogen cycle and plant uptake of nitrogen. (After Jansson. 1957)

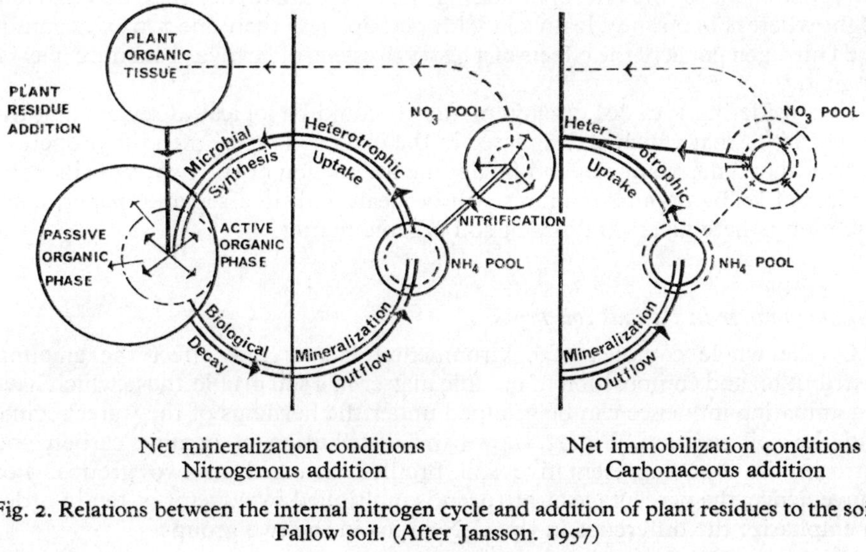

Fig. 2. Relations between the internal nitrogen cycle and addition of plant residues to the soil. Fallow soil. (After Jansson. 1957)

In these schemes two phases of organic matter are featured—the active and the passive. The former represents that portion of the carbon and nitrogen which circulates in the soil-plant system at a much more rapid turnover rate than the latter, which consists of more stable resistant compounds. Although the distinction between these two is arbitrary and diffuse, qualitively the active phase includes relatively fresh residues of the organic tissues of the soil flora, microflora and fauna, together with any readily decomposible organic manures which may be added to cultivated soils—corresponding approximately with the Nahrhumus defined by Scheffer and Ulrich[20].

'Priming Effect'

Broadbent and Norman[6] and Broadbent and Bartholomew[5], together with various other collaborators studied the mineralization of carbon and nitrogen when fresh green plant material labelled with C^{13} or C^{14}, and N^{15} was incubated with soil. By measuring the relative amounts of labelled and unlabelled CO_2 and NO_3 released they demonstrated that the surge of micro-organic activity increased the yield of both elements from the original humus in the soil. This phenomenon is generally termed the 'priming effect' and although it has been confirmed by other workers, both Jansson[14] and Harmsen[12] consider it to be of less practical significance than was at first supposed; in fact Jansson disputed the basis of the calculations made by Broadbent.

Types of organic matter

Factors considered in the introductory paragraphs suggest that the amounts of nitrogen released under field conditions are likely to vary considerably from soil to soil depending on environmental conditions for the micro-organisms and on the types and amounts of organic matter accumulated in or added to the soil. In a soil rich in organic matter containing ten tons per acre of accumulated nitrogen, or more, the effects of adding fresh organic matter may be relatively slight whereas in a sandy loam soil with perhaps less than one ton of accumulated nitrogen per acre the effects of a heavy dressing of farmyard manure may be dramatic.

In considering types of organic matter it would be logical to review first the properties of that which accumulated in the main soil types, then the properties of typical organic manures, and finally the interaction of the two; but since the subject of bulky organic manures will be dealt with in a separate paper, chief attention is here given to types of soil organic matter.

Organic matter in various soil types

Of the whole complex of environmental factors that affect the amount, distribution and composition of organic matter in a soil profile, those which have a dominating influence can be grouped under the headings of the water regime and the soil reaction. Fig. 3 shows the distribution of organic carbon and nitrogen in four representative soil profiles arranged in two groups. For convenience, the percentage of nitrogen is multiplied by a factor of ten in order to emphasize the difference in the C:N ratios in the two groups.

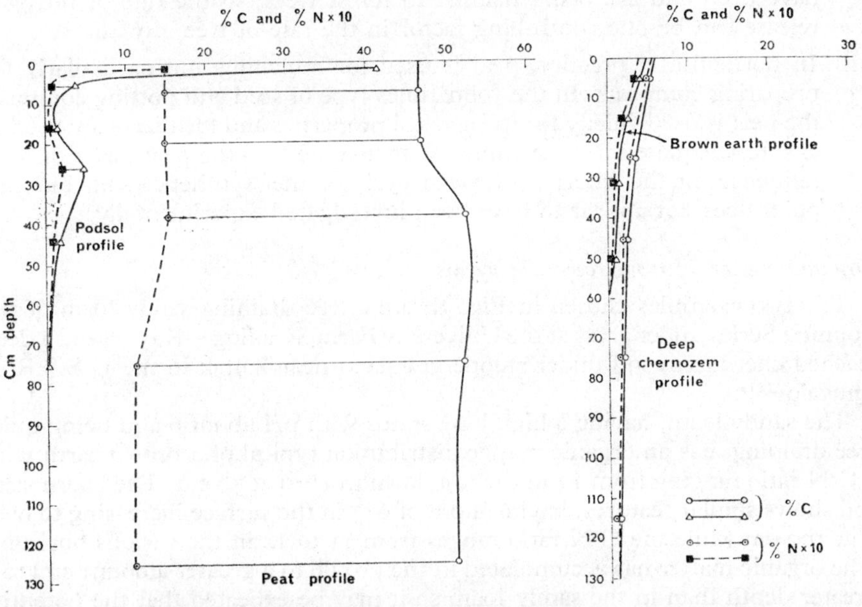

Fig. 3. Distribution of carbon and nitrogen in profiles

Organic matter with wide C:N ratios

The two examples chosen in Fig. 3a are a free draining humus podsol of the Crannymoor series under pines in Shropshire, (Soil Survey of England and Wales), and an acid basin peat from St. Fergus Moss, Aberdeenshire, composed principally of Eriophorum and Sphagnum spp., Glentworth and Muir[11].

The podsol has well defined L, F and H layers of organic matter with C:N ratios of about 30, overlying the mineral soil, and the B horizon clearly shows the accumulation of humus with a C:N ratio of 25 so that the whole of the organic matter has a relative excess of carbon and the release of nitrogen as ammonium can confidently be stated as small, although there is no direct experimental data for this particular soil. Under British conditions the total uptake of nitrogen by coniferous forest is about 40 lb per acre per annum, of which about 30 lb returns to the soil in the litter for recirculation.

The peat moss was sampled to a depth of 700 mm (20 ft) although the data presented in Fig. 3 extends only to 120 cm. In this section the C:N ratio ranges from 31 in the top 10 cm to about 50 at depth. Thus, the total amount of carbon and nitrogen in the profile is very great but the relative excess of carbon is also very large. These features are shown to different degrees by the shallower hill peats and by the neutral fen peats.

The nitrogen releasing properties of such peats are important from three standpoints:

(i) Agriculturally, when virgin peats are drained and brought under cultivation there is potentially a large reserve of nitrogen, but little may be released until the C:N ratio has been appreciably narrowed by aerobic decomposition and this is likely to occur more rapidly in fen peats with a high base status or after liming of the acid peats.

(ii) In forestry practice, considerable areas of peat in the wetter districts have been and are being planted to forest trees, so the rate of nitrogen release can be one controlling factor in the rate of tree growth.

(iii) In horticultural practice, peat is used for mulching and particularly for preparing composts. In the John Innes type of seed and potting composts the peat is used chiefly for its physical properties and fertilizers are used to ensure adequate nitrogen supplies. It may be that the peat has a distinct influence on the internal nitrogen cycle of the 'synthetic' soil, but this point does not appear to have been investigated experimentally.

Organic matter with narrow C:N ratios

The two examples chosen in Fig. 3b are a free draining sandy loam of the Sonning Series under grass at the University Farm, Reading—Kay[15] and a deep leached chernozem soil under steppe grassland near Kursk in the U.S.S.R.—Shuvalov[21].

The sandy loam, having a high base status with pH about 6 and being quite free draining, has an organic matter distribution typical of a brown earth with a C:N ratio ranging from 12 in the top down to 10·6 at 50 cm. The chernozem soil shows similar features, having a pH of 6·3 in the surface increasing to over 7 at 100 cm while the C:N ratio ranges from 11 to 12 in the various horizons. The organic matter has accumulated in the profile to a greater amount and to a greater depth than in the sandy loam so it may be expected that the potential nitrogen release will be greater, though the moisture content of the soil may not allow it to be realized fully during the growing season.

Relative nitrogen release

It is very difficult to secure comparable data for the release of nitrogen from different types of organic matter under the natural conditions in which they are found. From incubation data Scheffer and Ulrich have presented the following relative potential nitrogen mineralization capacities for various soils:

		mg N mineralized per gC
Peat soils	—	1·6
Forest soils (Ah)	—	2·6
Arable soils (sandy)	—	4·7
Various meadow soils	—	4·9
Pararendzinas	—	5·4
Arable soils	—	6·7

Chemical release of nitrogen forms from organic matter

Amino-acid components

The extensive studies initiated by Bremner[2,3], and extended by other workers such as Stephenson[22,23,24] in the U.S.A., and Kononova[16] in the U.S.S.R. have demonstrated that when soils are heated with 6N hydrochloric acid, or with sodium hydroxide under an atmosphere of nitrogen, a high proportion of the organic nitrogen is brought into solution as amino acids. These have been identified and determined by chromatographic techniques on paper or resin columns and the pattern of amino acids is found to be much the same for a wide range of soils.

The total quantity of amino nitrogen may be estimated by the Van Slyke reaction with nitrous acid or the α amino acid nitrogen determined more specifically by the colorimetric reaction with indanetrione (ninhydrin), or the yield of CO_2. Commonly it is found that between one third and half the total soil nitrogen is in the form of amino acids, and Bremner considers that these estimates are inclined to be low because of the deamination of the amino acids to produce ammonia which occurs during the hydrolysis procedure.

Amino-sugar components

Bremner and Shaw[4] and subsequent workers including Stephenson[21,23,24] have identified amino sugars, principally 2-amino glucose, in hydrochloric hydrolysates in amounts usually about 5–10 per cent of the total nitrogen. Parsons[18] studied the hydrolysis conditions and obtained somewhat higher yields with a mixture of hydrochloric and formic acids. Since amino sugars are not present to any significant extent in higher plant tissues that found in soils must be derived either from the cell walls of various species of bacteria and fungi which are known to contain amino sugars, or from the chitinous tissues of insects and other soil fauna.

Nucleotide components

Small quantities of purine and pyrimidine nitrogen compounds characteristic of nucleotides have been identified in soils but such studies as those of Adams[1], have been concerned mainly with the occurrence of nucleotide phosphorus rather than nitrogen.

Acid insoluble 'Humin' components

Commonly about one third of the total nitrogen remains insoluble when a soil is subjected to hydrolysis treatment with 6N hydrochloric acid at 100°C and, so far, little attempt has been made to elucidate its precise chemical composition. It is usually assumed that this fraction is virtually unavailable to plants and represents a core of insoluble resistant material with some of its nitrogen present as heterocyclic components in a complex condensed hetero-polymeric nucleus of the so called 'humic acid'. Much of the work of Flaig[10] and his associates in Germany has been devoted to the study of condensation reactions between quinone compounds related to lignin and lignin decomposition products, and amino compounds.

That the nitrogen of soil organic matter is not evenly distributed in the organic polymers is indicated by an extraction study withformic acid on a meadow soil made by the author. The results presented in Table 1 show that the organic material dissolved in the first two extracts is low in total nitrogen (2·15 and 3·36 per cent respectively) but nearly all this nitrogen is in the form of amino acids and amino sugars, whereas the material extracted later has a much higher proportion of chemically undefined nitrogen in its composition.

Table 1

Extraction of organic matter from meadow soil with boiling formic acid/acetylacetone/fluoroboric acid mixtures

20 g meadow soil boiled successively with 100 ml formic acid, 10 ml acetylacetone and 5 ml HBF_4

Extraction Series	Yield % of total (ash free O.M.)	Ash content %	Total N content % (ash-free)	α amino N as % total N %	Amino sugar N of total	α amino + amino sugar N
I 10 mins.	28·4	2·62	2·15	82·4	14·0	96·4
II 10 mins.	21·4	4·82	3·36	78·0	8·9	86·9
III 10 mins.	15·1	2·45	5·14	60·0	5·5	65·5
IV 10 mins.	11·5	6·15	5·76	60·8	6·3	67·1
V 10 mins.	10·0	11·86	4·10	53·2	5·6	58·8
Total yield	86·4					

Nitrogen forms in soil horizons

Part of the difficulty of relating crop response to soil nitrogen, as to other nutrients, is the differential distribution of the element in each soil profile.

Stephenson[22,23,24] and also Bremner[3] have studied the distribution of forms of nitrogen in selected profiles. Stephenson hydrolysed the soils with 6N HCl under pressure at 120°C for 24 hours and determined the ammonia in the hydrolysates by distillation with 1N NaOH. This treatment gives the yield of 'free' ammonia plus that split off from amino sugars so that a separate colorimetric estimation of the latter was made and the amount of amino sugar N subtracted to give the ammonia N. It was found that the nitrogen in the hydrolysate which did not distil under alkaline conditions was nearly all in the form of amino acids.

On this analytical basis acid insoluble 'humin' N generally decreased down the profile from between 20 and 40 per cent of the total in the surface horizon to 10 per cent or less at 50 to 100 cm depth. The proportion of amino acid nitrogen showed a similar trend, while the 'free' ammonia generally increased down the profile and the amino sugar varied between 3 and 24 per cent without showing any definite trends.

Chemical distribution of added nitrogen in soils

Using techniques which are basically an extension of the scheme used by Stephenson, recent American workers have attempted to study the distribution of nitrogen added to soils in the form of fertilizers. Cheng and Kurtz[7] found that most of that portion of the nitrogen added as ammonium sulphate to three silt loam soils which became incorporated in the organic matter was accounted for in the acid hydrolysis products as amino acid, amino sugar and other soluble nitrogen compounds.

Release of nitrogen by mild hydrolysis treatments

Although it is recognized that incubation techniques for the determination of mineralizable nitrogen probably offer the best prospects of a close correlation with the growth response of crops under field conditions, various workers have sought relatively mild chemical treatments as more convenient laboratory procedures. These include the determination of ammonia released by incubation of air dry soil at 28°C with N NaOH for 40–42 hours, which Cornfield[9] found was well correlated with microbiologically mineralized nitrogen and that formed by evaporation of a soil with $N/^{14}$ H_2SO_4 on a steambath according to the method proposed by Purves and Leo[19]. The former will release ammonia principally from amino sugars and the latter from less precise sources.

More recently MacLean[17] has reported a high correlation between the total nitrogen extraction by 0·01M $NaHCO_3$ solution on shaking for 15 minutes and the nitrogen uptake by oats, but in this case it is more likely that the bicarbonate solution merely dissolves readily mineralizable nitrogen compounds than releases ammonia by hydrolysis.

Conclusions

Assuming Jansson's idea of an 'active' portion of the soil organic matter cycling at a faster rate within the bulk of 'passive' material to be qualitatively correct, then both microbiological and chemical tests for available nitrogen should be designed to measure the critical parameters of this 'pool'.

Most microbiological tests, whether aerobic or anaerobic utilize the microflora to integrate the release of available mineral nitrogen from the pool of 'active' carbon and nitrogen compounds in the soil—admittedly under conditions which may be very different from those prevailing in the field—but the laboratory answer presumably represents the potential mineralization.

In the same way any purely chemical test for available nitrogen is unlikely to succeed widely unless it provides information on the amount of both the 'active' carbon and nitrogen.

References

1. ADAMS, A. P,. BARTHOLOMEW, W. V. and CLARK, F. E. (1954). *Soil Sci. Soc. Amer. Proc.* **18,** 40.
2. BREMNER, J. M. (1950). *Biochem. J.* **47,** 538.
3. BREMNER, J. M. (1958). *Rep. Rothamst. Exp. Sta.* 1957, 57.
4. BREMNER, J. M. and SHAW, K. (1954). *J. agric. Sci.* **44,** 152.
5. BROADBENT, F. E. and BARTHOLOMEW, W. V. (1948). *Soil Sci. Soc. Amer. Proc.* **13,** 271.
6. BROADBENT, F. E. and NORMAN, A. G. (1946). *Soil Sci. Soc. Amer. Proc.* **11,** 264.
7. CHENG, H. H. and KURTZ, L. T. (1963). *Soil Sci. Soc. Amer. Proc.* **27,** 312.
8. Commonwealth Bureau of Soils, (1964). *Collected Bibliography No.* 666.
9. CORNFIELD, A. H. (1960). *Nature,* **187,** 260.
10. FLAIG, W. (1964). 'Use of Isotopes in Soil Organic Matter Studies', F.A.O. Conference Report, **1,** 193.
11. GLENTWORTH, R. and MUIR, J. W. (1963). The Soils of the Country round Aberdeen, Inverurie and Fraserburgh. H.M.S.O. Edinburgh.
12. HARMSEN, G. W. (1964). Transactions of Eighth International Congress of Soil Science—in press.
13. HARMSEN, G. W. and VAN SCHREVEN, D. A. (1955). *Adv. Agron.* **7,** 299.
14. JANSSON, S. L. (1957). *Kungl. Lantbruks/Hogskolans,* **24,** 101.
15. KAY, K. F. (1936). Bulletin 49. University of Reading. 'A Soil Survey of the University Farm'.

16. KONONOVA, M. M. (1961). 'Soil Organic Matter', Pergamon Press, Oxford.
17. MACLEAN, A. A. (1964). *Nature.* **203,** 1307.
18. PARSONS, J. W. (1958). Thesis, University of Reading.
19. PURVIS, E. R. and LEO, M. W. M. (1961). *J. Agric. Food Chem.* **9,** 15.
20. SCHEFFER, F. and ULRICH, B. (1960). 'Humus und Humus dungung', Verlag, Stuttgart.
21. SCHUVALOV, S. A. (1964). Short Guide to Soil Excursion, Moscow,—Kherson.
22. STEPHENSON, F. (1957) a. *Soil Sci.* **83,** 113.
23. STEPHENSON, F. (1957) b. *Soil Sci.* **84,** 99.
24. STEPHENSON, F. (1957) c. *Soil Sci. Soc. Amer. Proc.* **21,** 283.

Discussion

Mr. T. G. Heafield in opening the discussion, supplemented the data presented by *Dr. Tinsley* by reference to long term plots established at the N.A.A.S. East Midland Regional centre at Shardlow.

The soil was of a sandy loam derived from deep gravels laid down at the confluence of the Rivers Trent and Derwent and typical of much of that water deposited material. Naturally acid, the soil was for a long period under parkland grass. The organic top soil had been removed prior to establishing the plots, resulting in a mineral soil with about 3·5 per cent organic matter.

On one block, a range of pH values from 4·0 to 7·0 had been established with standard (complete) fertilizer dressings; on a second block a range of fertilizer dressings were applied on plots of various pH values, whilst on a third block, the differing fertilizer dressings had been used on the soil of a constant pH value (6·5). A four course rotation of cereals, grass, roots and legumes had been established.

At all pH values, there had been a gradual reduction in total nitrogen values over the period of ten years from 1954–63. The mean values for total nitrogen in the soil had fallen from 0·16 per cent to 0·14 per cent. However, sampling in July 1964 had shown the following data for nitrate-N (p.p.m.):

		pH 4·5	pH 5·5	pH over 7
Potatoes	0–6 in.	7	5	7
	6–12 in.	8	6	5
Sugar Beet	0–6 in.	27	5	11
	6–12 in.	24	9	6

Sugar beet had failed at pH 4·5 whilst potatoes had grown satisfactorily at that pH value and this was reflected by higher nitrate levels in the 6–12 in. horizon as well as the 0–6 in. horizon.

On the block with soil pH 6·5, variation of fertilizer treatments (Nil; nitrogen only; phosphate and potash only; complete, nitrogen phosphate and potash) all showed a gradual reduction in total nitrogen in the soil. The significant feature of crops grown under these conditions was the continued moderate crop production from plots receiving no added fertilizer. Potassium rather than nitrogen seemed to be the limiting nutrient as potassium deficiency symptoms were common. Added nitrogenous fertilizer aggravated the deficiency symptoms and this was reflected by the soil nitrogen levels falling slightly more quickly.

The use of nitrogen with phosphate and potash appeared to limit the loss of soil nitrogen compared with where phosphate and potash only had been used. Where sugar beet and potatoes were grown after ten years of cropping with a nitrogen only treatment, potash deficiency precluded maximum growth and resulted in higher levels of soil nitrate-nitrogen in July 1964, both in the 0–6 in. horizon and the 6–12 in. horizon.

Other work has shown that the loss of nitrogen took place slowly year by year, more particularly where crop residues were removed completely, while the rate of loss was reduced if crop residues were returned to the soil. Any factor inhibiting maximum uptake by the crop, e.g., acidity or lack of other nutrients, appeared to accentuate the loss of soil nitrogen, presumably as the easily leached nitrate. Nitrate production appeared little affected by low soil pH values, but utilization was affected by the ability of different plants to tolerate acidity.

The use of a complete fertilizer appeared to ensure not only best crop yields but also slowed the fall of soil nitrogen levels.

Dr. T. H. Rose asked *Dr. Tinsley* if he thought the benefit of the high organic matter content of the Chernozen soils was mainly on soil structure in view of the large responses to nitrogen obtained on these soils. *Dr. Tinsley* said in reply that he did not think that the emphasis should be put on soil structure. He though the large response to nitrogen was due to limited release of nitrogen during a very short growing season.

Dr. D. S. Jenkinson commented on the processes of nitrogen mineralization when a grassland soil such as a chernozen was brought into cultivation. There were two distinct processes:– a fast process which dominated during the first few years and a later slow process. The fast process resulted in considerable release of nitrogen as nitrate, until the organic matter containing the readily mineralizable nitrogen had broken down when the slow process became dominant and resulted in a slow rate of nitrogen release. *Dr. Jenkinson* also said that the 'priming action' of fresh organic matter, resulting in accelerated decomposition of soil organic matter, had been greatly exaggerated in the earlier published work and recent work had suggested that the priming action was very small. The stabilizing of soil organic matter is due to a chemical stability and not to a lack of fresh organic matter, bacterial population or necessary enzymes. He thought that the soil organic matter consisted of at least three fractions rather than the two as suggested by Jansson. These are (1) an easily decomposed fraction with a half life of two to three years, (2) a less readily decomposed fraction with a half life of about 25 years and (3) a fraction almost completely inert.

Mr. W. Dermott asked if *Dr. Tinsley* would comment on the black sandy soils, which occur in several parts of the country, in which the sand grains appear to be completely covered by inert, black organic matter and on which manganese deficiency was common. *Dr. Tinsley* replied that he had no experience of these soils and did not know whether the organic matter would affect the availability of manganese.

Mr. L. J. Hooper asked *Mr. Heafield* if much work had been done on grass-clover swards. One might expect an accumulation of nitrogen on the PK treated plots. *Mr. Heafield* replied that grass-clover ley treatments had only been included for one year so it was not possible to draw any definite conclusions but there appeared to be a slight increase in soil nitrogen on the PK plots.

Mr. R. D. Russell asked if details of rainfall and the amount of water available for leaching of nitrate had been kept. *Mr. Heafield* replied that rainfall data was available but that no attempt had been made to correlate these with the soil nitrate levels; the soil on the plots was freely drained.

Prediction of nitrogen requirements of arable crops in mainly arable cropping systems

By G. W. COOKE, Ph.D. (Lond.), F.R.I.C.
Rothamsted Experimental Station

THE first section of these notes shows the need to improve the precision with which nitrogen fertilizers are applied. The amounts of nitrogen needed in different conditions are discussed in the second and third parts which are extracted from two published papers. Some of the results shown here are from experiments done by colleagues at Rothamsted which have not yet been published elsewhere.

Current use of nitrogen fertilizers

Changes in N used since 1944

The following figures, derived from Surveys of Fertilizer Practice, show how the rates of N fertilizers used have changed in the last 20 years:

Approximate overall average rates
(units N/acre)

	Cereals	Potatoes	Sugar beet	Grassland	
				temporary	permanent
1944	8	43	72	3	1
1950	12	70	90	9	5
1957	24	110	105	19	7
1962	42	125	123	38	17

Average actual rates in 1962 and percentages of fields dressed

	Wheat	Barley	Potatoes	Sugar beet	Grassland	
					temporary	permanent
Rate, units N/acre	60	44	128	123	56	44
per cent of area dressed	81	91	98	100	70	39

Optimum dressings

These rates may be contrasted with estimates of optimum dressings derived from recent experiments.

	units N/acre
Winter wheat	0—80
Spring wheat	60
Spring barley	30— 90
Potatoes	40—130
Sugar beet	60—120
Grass, for one cut only	60
Grass grazed by dairy cattle (total in the season)	150

The present problem

Twelve years ago the average overall use of nitrogen on cereals was less than 20 units N/acre, now the average use is twice as much and many individual farmers give much more than this. The average dressings now used on potatoes and beet are nearly as much as can be recommended for the most responsive crops. With all cash crops the nitrogen contributed by farmyard manure, by previous crops and by fertilizer dressings is ignored by most farmers; Surveys of Fertilizer Practice show that allowances for soil, rotation and season are not made systematically.

When farmers used much less than the optimum recommendations of N, lack of knowledge about how much should be used on a particular field did not greatly matter. But the success of propaganda to increase the use of N has produced the present problem; as average manuring approaches the published optima, it is important to know how to fit dressings to crop, soil, farming system and season. If too little N is used, yield will be lost, but if too much is given, yield may be less than the maximum, and any surplus is liable to be lost before another crop is grown. It is already important that nitrogen dressings should be related to local conditions for cash root crops and for cereals. Rapid increases in the last seven years in the amounts of nitrogen used on grass suggest that there may be inadequate experimental information for advising on the efficient use of N on grassland.

The present complacent approach to nitrogen manuring seems to come from accepting the good crops that are now common as the best that may be grown. Many recent experiments however, show that still higher yields are easily obtained; several cwt/acre more cereal grain, 2–3 tons/acre more potatoes, are often obtained by fitting fertilizer dressings closely to previous cropping and manuring instead of using average recommendations. These extra yields may not seem worth the trouble of getting them when farming is affluent; they cost little except a small amount for extra fertilizer, they may even be obtained by using less fertilizer than the manufacturer's leaflet suggests. They only involve applying existing knowledge on the effects of soil, cropping system and climate on the response to N; but this demands experience and thought which are both more scarce than money at present. Yet the still higher yields that are possible from correct fertilizing, and the savings from avoiding waste of fertilizer, could raise current profits; the improved technical efficiency so achieved could be very important in more competitive farming than exists at the present time.

Factors affecting returns from nitrogen

The important factors that affect return from nitrogen in arable farming are rotation, previous manuring and weather. Some aspects are discussed below.

1. *Previous crops complicate the effects of N fertilizers and the value of residues in soil*

Annual crops of grass often increase the need of following arable crops for fertilizer. In a recent experiment at Rothamsted by F. V. Widdowson and co-workers, concentrated fertilizers and four sources of N were tested on a rotation of kale, ryegrass, barley (all the crops were grown in any one year) from 1960–62; barley uniformly manured with 60 units N/acre was taken in 1963 and its yields were influenced both by the crop that preceded the barley and by the N manuring given in the previous three years.

Total N given in 1960–62 (units N/acre)	1960 1961 1962	kale ryegrass barley	ryegrass barley kale	barley kale ryegrass
			1963 BARLEY CROPS	
		severe lodging	moderate lodging	slight lodging
			Yields (cwt/acre)	
0		39	38	28
460		40	40	36
			percentage of N in grain	
0		1·8	1·7	1·5
460		1·8	1·8	1·7

In the presence of 60 units of fresh N/acre (usually regarded as nearly enough for barley) there were good yields where 1963 barley followed barley or kale, but the crop after ryegrass was much poorer. Residues from the large dressings of N given in 1960–62 had little effect where preceding crops were barley or kale but, after ryegrass, these residues went some way to remedying the serious deficiency of N caused by the grass.

2. *The preceding crop may be a source of nitrogen*

The large contributions that legumes may make to supplies of nitrogen often are not recognized. In a Rothamsted experiment by D. Slope, barley preceded by barley was compared with barley preceded by beans and several rates of N were tested; 1963 yields were:

N/acre to 1963 barley (units)	1962 crop: Barley	1962 crop: Beans
	Yields of 1963 barley, cwt/acre	
0	26	35
30	33	42
60	36	41
90	40	33

The yield without fresh N fertilizer was 9 cwt/acre more after beans than after barley. After beans 30 units N/acre was the right dressing but 90 units were needed after barley. If the nitrogen supplied by beans had been disregarded and a flat rate of 90 units N/acre given, there would have been a loss of nearly 10 cwt/acre of barley grain.

A single year of red clover can also increase soil N temporarily. At Rothamsted, F.V. Widdowson compared winter wheat taken after clover or after ryegrass; wheat after clover yielded 47 cwt., but after ryegrass only 31 cwt/acre. After clover 50 units N/acre gave a good increase in yield of wheat but there was only a small gain from using twice as much; after ryegrass the second increment gave a large increase in yield:

Units N/acre in 1962	1961 crops	
	Clover	Ryegrass
	Yields of 1962 wheat cwt/acre	
0	47	31
50	56	46
100	58	56

3. *Diseases transmitted by previous crops may affect the response to nitrogen*

Where an infection is transmitted to following crops both potential yield and response to N may be affected. In an experiment at Rothamsted, D. Slope measured yields of wheat grown in different rotations with low and high rates of N; 1961 yields in two of the rotations used were:

Rate of N to 1961 wheat units/acre	*Previous crops*	
	1958 Winter Wheat 1959 Winter Wheat 1960 Spring Wheat	Winter Wheat Oats Beans
	Yields of 1961 wheat, cwt/acre	
50	28	40
100	30	47

In the all-wheat rotation there was serious take-all infection, yields were poor and the second 50 units of N had only a small effect. There was little take-all after oats and beans, yields were much higher and, in spite of the extra nitrogen left by the beans, the high rate of fertilizer N was fully justified.

4. *Seasons affect responses to nitrogen*

It is known that weather affects the return from N in several ways:

(a) *Cyclic variations* in yield, with a period of 4–5 years, have been found in Holland by van der Paauw[32]. The responses of rye and potatoes to N were closely related and varied much through the cycle; average yields of wheat and rye were 50 per cent more after a sequence of dry than of wet years. These effects and their causes are too little understood for present information to be used in advisory work.

(b) *Summer rainfall* also influences yield and responses to nitrogen and other fertilizers as shown by van der Paauw[31]. But, as the weather to come cannot yet be forecast when crops are planted, there is little which can be done to make nitrogen manuring of arable crops fit the season more precisely.

(c) *Winter rainfall*, by determining whether leaching of nitrate is more or less severe than average, affects the amount of nitrogen available to a following crop.

Boyd, Garner and Haines[6] found half the variance in N responses by sugar beet was associated with winter rainfall. Similar experiments have been done by Dutch, German and American workers.

In Holland, van der Paauw showed that the effect was associated with the nitrogen lost from the soil rather than with the other effects of the extra rain. He found[32] that the effect of 1 mm of additional rain in a relatively dry winter was equalled by supplying an extra 0·25 to 1·25 kg N/ha (the actual amount depending on soil type). He concluded: 'The farmer may suffer great losses of nitrogen which are more or less inevitable and in addition—since he is unable to adjust for this—the depression of crop yield may be still more serious. Nor has he any idea of the amount of available nitrogen in the soil after dry winters. In both cases 'normal' dressings with nitrogen are unsatisfactory since they result in serious yield depressions due to deficiency after wet winters and increased possibilities of lodging causing loss of yield after dry ones. Fertilization adjusted to weather conditions in the preceding winter might contribute considerably to an increase of average yield and would also tend to an improved stabilization of agricultural production. For these reasons, in the Netherlands during the last three years general advice about the most profitable nitrogen dressing has been given to the farmers. For instance additional amounts up to 20 and 30 kg nitrogen per ha were recommended in 1959 and 1961 respectively for winter cereals, a reduction of 20 kg per ha was advised in 1960. ('Normal' dressings are about 70 to 80 kg/ha.) A noticeable increase of yield might be expected from the former advice without dangerously increasing the chance of lodging. In the latter case a decrease of lodging was to be expected without depressing the yield and with a saving of fertilizer costs. The experience obtained thus far appears to be encouraging.'

(d) *Leaching and denitrification*

For British conditions the actual importance of leaching in removing nitrate from soils is little understood. Heavy winter rain is believed to remove all the measurable nitrate from surface soil, but many recent experiments have shown that the large dressings of N now commonly used often leave large residues that benefit crops in the year following. Overseas experiments have shown that deep rooting crops recover nitrate that accumulates in subsoils; little is known of the possibilities of deep-rooting British crops 'mining' nitrate or whether there is much upward movement of nitrate in dry weather. In the drier parts of the country little drainage water passes through cropped land in summer. Temporary surface waterlogging may allow denitrification to occur, but there is little field evidence on the practical importance of this way of losing N from soil except the work of K. Shaw in North-east England[30].

Extract from:

'The Basis for Modern Manuring' by G. W. Cooke,
A lecture given to the Devon County Agricultural Association at Seale Hayne College.

Recent work with fertilizers on arable crops

Nitrogen for cereals

In the first comprehensive account of post-war experiments on cereals, E. R. Bullen and W. J. Lessells[10] showed that modern stiff-strawed varieties of cereals lessened the risk of loss by lodging and also gave larger returns per unit of nitrogen than did the older varieties. Up to 70 units N/acre were justified on new varieties of winter wheat and spring barley. Because modern varieties resist lodging, farmers can now use safely the 60 units of nitrogen per acre which seemed so dangerous when Crowther and Yates' work was published in 1941.

Many hundreds of experiments testing nitrogen on cereals have been done in the last 20 years and D. A. Boyd[7] showed that, while responses per unit N have remained surprisingly constant, yields of the crops grown without nitrogen (Table 1) were now much larger.

Table 1

Yields of cereals grown without nitrogen (cwt/acre)

Years of the experiments	Barley	Wheat
1900–40	18	20
1945–55		25 (all cereals)
1955–date	28	32

The most recent work, now being published by W. J. Lessells and J. Webber[23] shows that large differences in response to nitrogen are caused by differences in recent cropping, soil texture, climate and farming system.

Winter wheat grown without N on soils recently ploughed from grass gave average yields of about 36 cwt/acre; on land that had been mainly in arable crops yields were about 10 cwt/acre less. The response to 60 or 80 units N/acre on the 'old arable' soils was about three times greater than on soils that had recently been under grass. The wheat following grazed grassland, or clover, obtained more nitrogen from the soil than wheat grown after grass that had been often cut and removed. Lessells and Webber recommend no N on peaty fen soils, after frequent organic manuring, or in high rainfall areas. On land recently ploughed from grass, they recommend 40 units of N for wheat after grazed grass and 60 units after grass cut for hay or silage. After arable cropping they recommend 60 units on heavy soils and 80 units on other soils.

Spring barley gave very variable results on sites recently ploughed from grass; responses to N were often small but occasionally they were very large. Response decreased as the level of yield without N went above 30 cwt/acre; with yields of over 40 cwt/acre without N, fertilizer N had little effect on yield. For barley grown on peat soils, in wet areas, or late sown, or after well-manured kale—no more than 30 units N/acre were needed. After ploughing from grass, 30 units N/acre were also recommended unless barley was known to respond well on the particular land. For barley grown after arable crops 60 units of N combine-drilled or 80–90 units per acre broadcast on the seedbed were recommended.

Spring wheat needed 60 units N/acre irrespective of previous cropping.

Combine-drilled nitrogen for spring cereals has been fully discussed recently[15]. Lessells and Webber[23] found that drilled N was most superior to broadcasting in barley experiments where yields without N were less than 25 cwt/acre. Where more than 35 cwt/acre was obtained without N, yields from broadcast dressings were at least equal to those from equivalent drilled dressings. For wheat there were no consistent differences between the two methods.

Potatoes

Fertilizer responses in potato experiments carried out since 1945 have been surprisingly similar to those in earlier work, but interactions between nutrients have been larger and more consistent than had been expected. D. A. Boyd[7] showed that response of potatoes (and sugar beet) to a particular nutrient increased with increasing level of 'basal' nutrients, he suggested that fertilizer dressings much larger than those previously tested might prove profitable. In the most recent experiments, which have tested large dressings, response curves did not rise to a maximum and then become flat (an exponential curve) but were parabolic, yields rising to a maximum and then falling off again as fertilizer rates increased. These effects were particularly striking with nitrogen, an excess depressed yields, particularly when too little phosphate and potash was applied. A few recent experiments where potatoes needed little phosphate seem not to have been typical and Boyd estimated[7] that the usual optimum is still at least 100 units P_2O_5/acre.

N. H. Pizer and others[28] found that, on the peat fenland optimal fertilizer dressings for nitrogen and phosphate were well above the largest tested in the experiments. Their recommendations ranged from 100 to 120 units N/acre and from 150 to 180 units P_2O_5/acre; the potash needed ranged from none on the richer organic silty loams to 200 units K_2O/acre on peats.

D. A. Boyd and W. Dermott[9] described 124 experiments done by the N.A.A.S. from 1955 to 1961. In spite of large changes in manuring and methods of potato growing, average response to phosphate was bigger than in the experiments done from 1900–40, average response to nitrogen was a little less and to potash much less. The following figures show how the optimum nitrogen dressings depended on cropping, soil texture and rainfall:

		Light soils	Heavy soils	Wet areas
			units N/acre	
After	Cereals	100	80	70
	Roots or ley	80	40	40 (?)

(The direct effect of high rainfall on nitrogen response, and its indirect effect through soil properties or farming type could not be distinguished.) Phosphate gave large increases in the west and northwest, and on all the heavier soils classified by analysis as 'low' or 'medium' in P; average responses were much less on sands and sandy loams where soluble P in soil was often 'satisfactory'. Boyd and Dermott suggested that potatoes on heavy soils should have 100–150 units P_2O_5/acre; the following optimum dressings (in units P_2O_5/acre) were suggested for light soils to allow for soil phosphate level: 'low' P soils 100 to 150, soils 'satisfactory' in P, 50 to 100, and on soils 'high' in P from none to 50 units. The optimum potash dressing was about two hundred units K_2O/acre on sands and sandy loams; on heavier soils there was a range from none to 100 units K_2O/acre.

Similar results were obtained in 50 experiments done by Levington Research Station[5] from 1956–62; most profitable average rates were 100 units N, at least 100 units P_2O_5, and 160 units K_2O. Response to nitrogen differed much from season to season and with previous cropping; optimum dressings in 'arable' rotations (after four tillage crops) were 130 units; in 'mixed' rotations (not more than one year after a short ley or three years from long ley or old grass) only 80 units were needed. Yields were depressed seriously by too much nitrogen used in ley rotations. There was less risk of losing yield by using excess nitrogen when potatoes followed tillage crops, or by giving too much phosphate or potash. In some later experiments by workers at Levington, responses increased up to 165 units P_2O_5/acre. At the highest level of phosphate tested, profitable increases in yield were still obtained irrespective of soil analysis. Sites where potatoes showed little or no response to phosphate could not be identified by analysis or type of soil or locality. The average optimum dressing of potash was 160 units K_2O/acre and there was a fairly good correlation with available soil potassium level. 200–250 units K_2O/acre were recommended on soils 'low' in soluble K, 160 units K_2O on 'medium' soils and 80–100 units on those 'high' in K.

Boyd[7] shows that 'balance' in NPK fertilizers is important for potatoes over a wide range of conditions and that, except on particularly rich sites, substantial losses can result from using fertilizers rich in nitrogen and potash, but deficient in phosphate. Potatoes have a special need for fresh dressings of fertilizer phosphate; in Rothamsted work they have responded well to fresh dressings where other crops have obtained enough phosphate from soil. It seems false economy to try to save phosphate for potatoes on soils with 'medium' levels; the subsidized cost of one cwt of superphosphate is about 10 shillings, and this is less than the value of a one cwt response which could not be measured accurately in field experiments.

Sugar beet

In experiments done between 1934 and 1949[6] 5 cwt/acre of agricultural salt gave large responses in almost all parts of the country and lessened the response to potash. Average optimum dressings, applicable to beet grown without salt in most regions, were 100–120 units N and 150 to 200 units K_2O/acre. Returns from phosphate were small, scarcely paying the cost of the dressings tested, and no more than 50 units P_2O_5/acre could be recommended. Responses to all nutrients were lessened in fields given FYM and response to nitrogen was much less on all fields that had been under a long ley or old grass within five years of the experimental beet crop. Response to nitrogen was also affected by winter rainfall, being greater than average after winters wetter than average and less after dry winters.

Later experiments by S. N. Adams[1] in 1957–60 indicated that average optimum dressings for beet grown without salt were:

	N	P_2O_5	K_2O
	units/acre		
Without FYM	100	50	160
With FYM	60	0	80
After a legume or heavily manured roots	60	50	160

Response to nitrogen differed greatly from field to field and the best guide to the correct nitrogen dressing was previous cropping. Beet grown after a legume or heavily manured roots needed only 60 units N/acre, after two or more cereals 120 units N were justified. Exceeding the optimum dressing of N depressed yields of sugar although it increased the yield of tops. Responses to phosphate and potash did not differ much in different experiments and the national optimum would have been satisfactory on nearly all fields; perhaps this is because most beet is grown in the east of the country on light soils under relatively standard conditions. Adams' work confirmed the value of 3–4 cwt of salt/acre; this dressing increased yields more than potash did. When salt was used no more than half the above amounts of potash were justified.

Swedes

Swedes are now mostly grown in the west and north of England and in Scotland. In experiments in north-east Scotland(29) done from 1948 to 1958 an exponential curve described the response to phosphate but, as with other crops, the curves for response to nitrogen and potash rose to a maximum and then fell again as dressings increased. Crop rotation affected nitrogen reponse; on farms where leys were grown for only one or two years up to 90 units N/acre were needed, but on most fields in this area leys are longer and 60–65 units of N were enough. Farmyard manure lessened fertilizer responses. Recommendations for the area were:

	N	P_2O_5	K_2O
	units/acre		
Without FYM	60	110	110
With FYM	45	100	50

The present basis for using fertilizers

Table 2 indicates how adequate is the information from experiments on how fertilizer should be used.

Nitrogen

Adequate advice for average conditions on using nitrogen for all of the important arable crops and for grass can be given. Most farmers' practices in grassland manuring lag far behind current-knowledge, and, as grassland manuring increases, more information will be needed on the response to nitrogen of different types of herbage used in different ways.

Table 2

Recent experimental work on the value of fertilizers

+ information adequate } for advice on
− information inadequate } average conditions

	for N	for P	for K
Potatoes	+	+	+
Sugar beet	+	+	+
Other row crops:			
Kale	+	−	−
Turnips	+	+	+
Mangolds	−	−	−
Cereals	+	−	−
Grass	+	−	−

How useful are average optimum dressings?

Recommendations for 'optimum dressings' apply at best to the average of conditions covered by the experiments on which the advice is based. They do not apply automatically to any particular field and, if manuring is to be efficient, recommendations for average conditions must be adjusted to allow for departures from average caused by the residual effects of previous cropping and manuring, levels of other nutrients, soil type and climate, diseases, and the way the crop is grown. When farmers used much less than the published optimum dressings, as happened 25 years ago, the optima were useful targets to increase fertilizer use. The present position is quite different; farmers use, on average, as much or more fertilizer as is generally recommended for many arable crops. Only for grass and a few arable fodder crops does manuring still lag seriously behind what we believe to be justified.

For heavily manured arable crops, advice based on average optima is inappropriate for many fields; often by giving these amounts fertilizer would be wasted, and sometimes yield would be lost. Average recommendations are best regarded as only a starting point for advice, which must be modified to allow for the following factors which are discussed in the rest of this paper:

(1) *Interaction effects* associated with methods and intensity of farming. (Interactions that are unimportant when yields are small may become very important when potential yield is increased by applying other nutrients or water, or by growing the crop differently.)

(2) *Past manuring* with FYM and fertilizers leaves residues of nitrogen phosphate and potash in the soil that benefit following crops. The residues of inorganic N fertilizers usually last for no more than a season; but residual effects of continued manuring with phosphate and potash may last for many years.

(3) *Farming system:* Crop rotations, methods of using crops, systems of keeping stock, and the use of farmyard manure all affect the supply of nutrients to crops.

(4) *Weather* affects fertilizer responses. No allowance can yet be made for the season that is to come, but the affect of winter rain on the nitrogen needed by crops can be allowed for.

Interactions with fertilizer effects

Nutrient interactions

Post-war experiments showed that interactions between nutrients become increasingly important as the amount of fertilizer given increases. Experiments with potatoes and sugar beet show that as the basal manuring with PK fertilizer for potatoes, or potash for beet, is increased, responses to nitrogen increase. Often large amounts of nitrogen are justified only when large amounts of PK fertilizer or potash are also given. Too much fertilizer-N lessens yields of roots, particularly when dressings of other fertilizers are inadequate. In one example on potatoes quoted by Boyd[7] with a low rate of phosphate, 80 units N/acre gave maximum yield and using 120 units N/acre gave about 0·5 ton/acre less. When twice as much phosphate was given, more than 80 units N/acre were justified and 120 units had no adverse effects.

When nitrogen is lacking yields of grass are affected little by potash manuring, but as Table 3 from Holmes and Maclusky[21] shows, when adequate nitrogen is available, full yields may not be obtained unless potash is also applied.

Table 3

Effects of potash and nitrogen on yields of grass cut continuously

Units N/acre/ year	Without K_2O	With K_2O
	cwt of dry matter/acre	
None	36	42
230	50	72
460	46	86

Interactions of nutrients and water supply

Irrigation and N-fertilizer dressing often interact. In some experiments water and nitrogen were roughly interchangeable; others showed the return from nitrogen was much increased by watering.

Interactions with methods of growing crops

When more plants grow on an acre, more nutrients are needed to produce the extra potential yields. American workers have shown that the optimal dressings of nitrogen for close-planted stands of hybrid maize are larger than for crops with smaller plant populations. In Britain some farmers are now sowing cereals in rows 4 in. apart instead of the usual 7 in. and more nitrogen can profitably be used with the narrow than with the wide rows.

Residues from fertilizers and manures

Tables of residual values used for tenant-right compensation assume that inorganicnitrogen fertilizers have no residual value, that inorganic phosphate leaves residues for three crops after the first and nothing afterwards, and that potash is exhausted after three crops are grown. Recent research shows that none of these assumptions is always correct. Inorganic nitrogenous fertilizers do leave residues; the residues of phosphate manuring may persist in the soil for many years and have increased crops half a century after the fertilizers were applied; some experiments show potash fertilizer residues also last for much more than three years but in other experiments single crops of kale or grass have removed nearly all of a potash fertilizer dressing.

Nitrogen fertilizer

Very old nitrogenous residues from organic manuring, or from periods under grass, are usually worth little[35] and may be ignored as sources of nitrogen for crops. Recent dressings of organic manures, or recent periods under herbage crops, leave much more active residues that cannot be ignored in planning manuring. The heavy dressings of inorganic nitrogen fertilizers now commonly used often leave residues that benefit following crops. F. V. Widdowson and A. Penny[36] found that 150 units N/acre given to potatoes in 1961 raised yields of the 1962 wheat, which received no fresh fertilizer nitrogen, from 28 to 37 cwt/acre; but even where the wheat received a fresh dressing of 100 units N/acre, residues from the dressing given to the previous potatoes still raised yields by 6 cwt/acre, suggesting that soil containing residues of inorganic nitrogen can give larger yields than are possible from soil without residues, however much fresh fertilizer-N is given.

The effect of farming system, crop rotation and manuring on nutrient supplies in soil

The main effects of farming system and rotation are on nitrogen and potash reserves. If modern 'average optimum' dressings of phosphate are applied, phosphate inevitably accumulates in soil. Even with cash cropping systems, in which less phosphate is given than crops remove, soluble phosphate is lost slowly and there do not seem to be any experiments that have run for long enough to convert a soil with satisfactory levels of phosphate into phosphate deficient soil.

Nitrogen

The percentage of nitrogen in soil under continuous arable cropping tends to a constant value that depends on soil type and climate as well as cropping and manuring. 'Standard' nitrogen fertilizer recommendations refer to such 'base-line' soils, their rate of nitrogen turnover is roughly constant and the amount of nitrogen released each year is far too little for successful crops in Britain. Old grassland soils contain much more nitrogen and when they are first cultivated much is released, and often in the early years after ploughing the nitrogen supplied is enough for maximum yields of arable crops. Nitrogen also accumulates in soil where leguminous green manures, arable legumes, or leys are grown; an arable crop that follows needs less applied nitrogen than the same crop in rotations without legume or ley. Adjustments to standard nitrogen manuring should always be made to allow for such effects.

Green manures: At Woburn, green manures have consistently increased yields of both barley and sugar beet, with trefoil more effective than ryegrass[2]. Green manures increased the general level of yield at Woburn; for example 100 units of fresh nitrogen per acre produced most sugar whether or not green manures were grown, but when trefoil was undersown in the barley that preceded the beet an extra 8 cwt/acre of sugar was produced, the gain after undersowing with ryegrass was only 2 cwt. Experiments at other centres have not, however, given such large benefits from green manuring.

Annual leguminous crops: Single-year crops of red clover can fix up to 250 lb N/acre; the results of a two-year experiment at Rothamsted (Table 4) show they are an outstanding preparation for wheat.

Table 4

Effects of preceding crops on wheat yields

Fresh nitrogen given to wheat (units/acre)	Preceding crop: Clover	Preceding crop: Ryegrass (+200 units N)
	Yields of grain (cwt/acre)	
0	49	35
50	56	46
100	58	54

The residues of a good clover crop, as compared with residues of ryegrass heavily manured with nitrogen, were worth about 55 units N/acre in terms of a fresh nitrogen dressing given to the wheat[37]. Few experiments have been done to measure the nitrogen residues left by beans and peas; work at Rothamsted in 1963 showed that maximum yield of barley grown after beans was obtained with 30 units N/acre, and when more was applied yield was lessened. In a continuous barley system 90 units N did not produce as much grain as 30 units N given to the crop that followed beans. Although yield of the continuous barley may have been lessened by soil-borne disease, the value of the beans for supplying nitrogen was clear.

Longer leys: Increases in the nitrogen available to arable crops following leys have been measured in many experiments at Research Institutes and on N.A.A.S. Experimental Farms. Striking benefits from a period under lucerne are common; typical results at Rothamsted are shown in Table 5 which compared average wheat yields (over the years 1961–63) after three years of lucerne with yields after a sequence of arable crops[8].

Table 5

Effects of previous cropping on the response of wheat to nitrogen

N given to wheat (units/acre)	Crops for 3 preceding years: Lucerne ley	Crops for 3 preceding years: Arable crops
	cwt/acre of wheat	
0	45	28
40	53	39
80	54	48
120	53	50

The lucerne left so much nitrogen in the soil that with only 40 units of fertilizer-N/acre, wheat yielded within a hundredweight of the maximum in the lucerne rotation; after arable cropping 120 units N/acre were needed for maximum yield. Soil enriched with nitrogen by the lucerne gave several hundredweights more grain than the poorer soil liberally treated with fresh N-fertilizer.

Many other experiments in England and Scotland show that legumes or leys in the rotation lessen the need for nitrogen fertilizer, commonly only a third as much N is needed as is justified in arable rotations. Often when too much N has been given to cereals following legumes, yields have been depressed. *Old grassland:* Often when old grassland is ploughed so much nitrogen is released that no fertilizer-N is needed; there is however little information on how long such effects last. Part of Highfield at Rothamsted was ploughed from very old grass in 1948; at first, crops grown on the field did not respond to N, but since 1960 responses have been similar to those measured on old arable land. Therefore fifteen years after ploughing century-old grass, the amount of nitrogen released is little greater than that from old arable land although the old grassland soil still has 2 per cent more organic matter, and is in better physical condition than the old arable soil.

Adjusting nitrogen dressings to season

Although season affects responses to all plant nutrients, the effect on nitrogen is the only one that can be allowed for. No laboratory method of measuring the amount that a particular soil will supply to crops in the coming season is reliable enough for general use in advising on manuring. (The best guide to the amount of fertilizer N needed is from knowledge of the soil and of previous cropping and manuring.) Heavy fertilizer dressings leave residues of nitrate in the soil, more is released as crop residues decay. These reserves of nitrate are easily leached, and the residue left to benefit a following crop depends on the amount of rain in winter. Experiments in both England and Holland have shown that after wet winters more than average N fertilizer dressings are needed; after winters that were drier than usual, soil has a bigger reserve of nitrate and crops need less fertilizer. During the last few years advice on manuring to farmers in the Netherlands has taken account of winter rain. For winter cereals 20 to 30 units of extra N/acre are recommended after wet winters, 20 units N/acre less are needed after a dry winter. ('Normal' dressings in Holland are 70–80 units N/acre.) As experience with this scheme in Holland is said to be encouraging, should British Advisory Services also revise their N-fertilizer recommendations each spring to allow for the previous winter's rain? Experiments carried out here have given enough data to provide a basis.

Spring and summer rainfall, and temperature, affect crop responses to N-fertilizers but, as they cannot yet be forecast accurately when spring fertilizer dressings are given, plans must usually be made assuming that the season will be 'average'. Extra N can be given in mid-season in very wet years to some crops, but in dry seasons good responses to dressings already given may not be obtained without irrigation.

Extract from:

'NITROGEN FERTILIZERS: Their place in food production, the forms which are made and their efficiencies'

by G. W. Cooke

A lecture given to The Fertilizer Society and published as Proceedings No. 80, 1964

Matching soil and fertilizer nitrogen to the needs of crops

Using nitrogen fertilizers properly means that two sources of supply—soil-N and fertilizer-N are fitted together to match the needs of the crop. Although more research is needed on the nature of soil nitrogen reserves and their value, too few farmers make the adjustments to manuring that are justified by present knowledge of differences in nitrogen supply from different soils. The bulk of the nitrogen in soil is old, being part of the humus; it is released very slowly during the turnover of soil organic matter (a process that is imperfectly understood). In addition soils contain variable amounts of a much more active nitrogen fraction which is released (as nitrate) more quickly. This nitrogen may simply be the residues of unused inorganic fertilizer dressings, but generally it is derived from recently added crop residues or organic manures; although only a small part of the total soil N, it often determines the nitrogen-supplying power of soil, and therefore how much fertilizer is needed to supplement it.

Estimating available soil nitrogen

Many laboratories have developed methods, usually involving biological incubation, of assessing the amount of inorganic nitrogen that soils may supply to crops. In a typical method J. K. R. Gasser[19] found (Table 6) that mineral-N measured in fresh soils, and the quantities of N mineralized by these soils when incubated, were significantly correlated with yields of barley grown without nitrogenous fertilizer, and also with responses to fertilizer-N.

Table 6

Average yields of barley on soils of different nitrogen status

Laboratory measure of mineral-N in fresh soil	Unfertilized yields of barley cwt/acre	Increase in yield from 56 lb N/acre cwt/acre	N in barley grain per cent
Low	20	14	1·55
Medium	28	6	1·69
High	36	−2	1·87

Corresponding relationships in a series of potato experiments were much less satisfactory. Under some restricted conditions, laboratory methods give data for 'available' N that correlate well with responses to fertilizer; but the methods seem too uncertain to be applied to the wide range of conditions met in advisory work. (A method recently developed by D. S. Jenkinson[22], which involves only extraction by $Ba(OH)_2$ solution, seems more promising.) For the present, advisers have to use their knowledge of the history of the land and the results of fertilizer tests on it, the crop rotation, previous manuring, and weather to guide farmers in their use of N on arable crops. Some factors that affect the amount of nitrogen that soils can supply are discussed below.

Residues from inorganic nitrogen fertilizers

It has often been assumed that inorganic nitrogenous fertilizers have no residual effect, but recent experiments show that the heavy dressings now commonly used may increase crops that follow, at least on heavier soils. In one set of experiments[36] at Rothamsted, 100 lb N/acre (as ammonium sulphate) used on potatoes had a residual effect for following barley equivalent to 17 lb/acre of fresh N as fertilizer; on sandy loam at Woburn, N had smaller, but still measurable residual effects. In other Rothamsted work, F. V. Widdowson and A. Penny[35] found that 112 lb N/acre given to potatoes in 1960 was equivalent to a fresh dressing of 28 lb N/acre for the 1961 wheat crop although the intervening winter was very wet. 112 lb N/acre given to wheat in 1960 had a residual effect on following potatoes equal to 22 lb of fresh N as fertilizer. 168 lb N/acre given to potatoes in 1961 raised yields of the 1962 wheat, which received no fresh fertilizer N, from 28 to 37 cwt/acre; even where the wheat received a fresh dressing of 112 lb N/acre, residues from the dressing given to the previous potatoes still raised yields by 6 cwt/acre. A. C. Owers[26] states that the practice of over-manuring for potatoes on silt soils, which still persists, leaves a surplus of fertilizer for following wheat. At Terrington Experimental Husbandry Farm, maximum yields of Cappelle wheat on heavy silt soils were obtained without using any N fertilizer where crops followed heavily-manured potatoes. Light silts retained less N than the heavy silts and dressings of 45 lb N/acre were justified for wheat after potatoes.

More work is obviously needed on the residual effects of N fertilizers on other soils, but the results discussed have interesting implications. Can the residual values of nitrogen fertilizers adopted for tenant-right valuation purposes still be regarded as correct? (they state that inorganic N fertilizers have no residual effect). Should allowance always be made for the nitrogen manuring of previous arable crops? At Rothamsted it seems possible to build up a temporary reserve of inorganic nitrogen in soil; with such reserves, yields have been higher than on soils without nitrogen residues, however much fresh fertilizer-N was given.

Residues of one-year leys and green manure crops

F. V. Widdowson *et al*[37] found that yields of wheat following red clover were consistently greater than those after Italian ryegrass manured heavily with N. Clover was an outstanding preparation for wheat; one-year crops that were cut and removed left residues worth 16 lb and 63 lb N/acre respectively (expressed in terms of a fresh spring dressing of N to the wheat) from a poor clover crop in 1960 and a good one in 1961. In 1962 wheat grown without N-fertilizer yielded 47 cwt/acre after good clover, but only 31 cwt after ryegrass; heavy dressings of N given to ryegrass left no worthwhile residues in the soil.

In experiments at Woburn[3], trefoil and ryegrass grown as green manures after early potatoes increased barley yields, trefoil much more than ryegrass.

When a low dressing of fertilizer-N (26 lb/acre) was given to barley it yielded 10 cwt more after trefoil than where it followed a fallow. Both trefoil and rye-grass undersown in barley raised yields of the early potatoes that followed by about 10 cwt/acre as compared with no green manure. G. V. Dyke[16] has reported another series of two-year experiments at Woburn where green manures under-sown in barley were tested on following sugar beet. Average yields over two years of the experiments were 57 cwt of sugar/acre without undersown trefoil and 65 cwt/acre after trefoil, both in the presence of 112 lb of fertilizer N/acre. Ryegrass increased yields but was less effective than trefoil. N fertilizer for the beet was also tested at rates well above those needed for maximum yield; undersowing the barley with trefoil increased yields of beet by 7–8 cwt of sugar/acre as compared with no green manuring at each level of fertilizer-N. In this recent work on the poor Woburn soil, green manures have increased yields above the levels that can be achieved by giving nitrogenous fertilizer in the usual way at 'optimum' rates. How these effects are obtained is not clear but it seems that nitrogen in residues of green manures can be more valuable than fertilizer-N.

Residues of longer leys

Most English evidence on increases in 'available' soil nitrogen caused by leys is from work now being done on the experimental farms of Research Institutes and the N.A.A.S.; results so far available have been discussed recently[14]. Most experiments show striking benefits from a period under lucerne; the crops that have followed lucerne have usually yielded more than the same crops grown after an arable sequence or after a grass ley. Typical results are the wheat yields obtained at Rothamsted in 1961–63 on Fosters Field (an old arable field) and shown in Table 7.

Table 7

Yields of wheat grown with several rates of nitrogen following lucerne or arable crops

N applied to wheat lb/acre	Preceding crops, 3 years of	
	Lucerne	Arable crops
	Wheat yields (grain, cwt/acre)	
0	45·3	28·4
45	53·1	39·4
90	54·2	47·7
135	52·6	50·1

The lucerne left much more nitrogen in the soil than arable crops did and the wheat that followed it, when grown without N-fertilizer, yielded 17 cwt more than after arable crops. Wheat receiving only 45 lb N/acre yielded within a hundredweight of the maximum in the lucerne rotation, but after arable cropping 135 lb N/acre was needed for maximum yield; the shape of the response curve suggested that even higher dressings of N would have given still higher yields in the arable rotation. The advantage from growing lucerne or other leys does not last long. In these experiments barley is the third crop after ploughing the leys; although it yields better after leys than in arable rotation, particularly when no fertilizer-N is given, the gains are proportionately less than for wheat.

Most experiments of this kind show that legumes grown alone, or grazed leguminous leys, leave residues in the soil that supply nitrogen and lessen the amount of fertilizer-N needed for maximum yields. At Jealott's Hill Low, Piper and Roberts[25] found that soil N mineralized for the two arable test crops of the rotation was from 90 lb N/acre in an all-arable rotation, to 400 lb N/acre after 3-year leys, this high release of nitrogen after leys was at least partly responsible for better yields of kale and wheat that followed the grass. At Hurley, C. R. Clement[12] found that wheat in an all-arable rotation obtained 50 lb N/acre from the soil, after three years of grazed ley it obtained 80–100 lb. In two Hurley experiments about 80 per cent of the variation in yields of wheat in different rotations was associated with nitrogen supply. But Clement is doubtful whether, with increasing land values and decreasing fertilizer prices, producing an organic source of nitrogen by leys is worthwhile on most arable farms. In south-east Scotland, Holmes *et al*[20] found that most fertilizer-N was needed by cereals on farms with all arable rotations and least on farms where leys occupied half of the rotation. The mean responses expected to 56 lb N/acre were about 2 cwt of barley grain in arable-ley rotation, $3\frac{1}{2}$ cwt on 'mainly arable farms' and over 6 cwt on all-arable farms. They considered 100 lb N/acre was likely to be economic for barley on intensive all-arable farms; on farms with a grazing ley in the rotation the 'optimum' was about 28–45 lb N/acre, this being reduced to none when the barley followed ley or crops eaten off by stock.

Few American experiments on ley and arable rotations have been done, but Carlson *et al*[11] give results confirming British work. They found maximum yields of arable crops grown in rotation with a three-year crop of lucerne were higher than in rotation with non-legumes. Yields of both maize and barley that followed the lucerne were larger than after other crops, however much fresh N-fertilizer was given.

Leguminous leys and green manures leave supplies of nitrogen in soil that are useful to following crops, but in addition these residues often increase crop yields in ways that cannot be easily imitated by N-fertilizer dressings broadcast in the ordinary way on the soil surface and harrowed in. (Perhaps this is because the residues are distributed through the whole surface layers of the soil.)

Interactions of rainfall with nitrogen

Allowing for the effects of winter rain

Winter rainfall, by affecting the amount of nitrate leached from soil, influences the fertilizer nitrogen needed by following crops. Long ago R. A. Fisher[17] found a negative correlation between winter rain and yield of wheat at Rothamsted. Boyd *et al*[6] found that 1 in. above (or below) the mean rainfall per winter month increased (or decreased) the response of sugar beet to N-fertilizer by $2\frac{1}{2}$ cwt of sugar/acre. Recently F. van der Paauw[32] in Holland found a close negative relationship between the sum of November to February rainfall and the yield of rye grown without N-fertilizer. Over the range of soils investigated 1 mm of additional rain in winter meant that 0·25 to 1·25 kg of extra N/ha was needed in the following spring. Van der Paauw considered that spring N-fertilization should be adjusted to allow for the effect of the preceding winter's weather on soil nitrogen reserves. During the last few years general advice on the most profitable spring nitrogen dressing given to farmers in the Netherlands has taken account of winter rain. For winter cereals additional amounts of 20 and 30 kg N/ha. were recommended respectively after the wet winters of 1959 and 1961. A reduction of 20 kg N/ha was advised in 1960. ('Normal' dressings are 70–80 kg N/ha.) Experience with this scheme in Holland is said to be encouraging. Should British Advisory Services also revise their N-fertilizer recommendations each spring to allow for the effects of the previous winter's rain on the N that soils will supply?

The effect of spring and summer rain

Rainfall and temperature both affect crop responses to nitrogen fertilizers but, as neither can be forecast accurately when spring fertilizer dressings are given, plans must usually be made by assuming that the season will be 'average'. Extra N can be given in mid-season in very wet years to some crops, but in dry seasons good responses to dressings already given in spring may not be obtained unless irrigation water can be supplied.

Irrigation and nitrogen fertilizer

Conifer seedlings

B. Benzian's experiments[4] on Sitka spruce show well the benefit of watering in dry spells. In the dry summer of 1947 nitrogen was valueless, or slightly harmful where seedbeds received no water, but increased heights on watered plots. With water, the gains from N applied in July were about twice those from June or August dressings. Water had no effect on plants that received no nitrogen.

Grass

Relationships between the use of nitrogen fertilizer and irrigation on grassland have been discussed[13]. Experiments at Jealott's Hill[24] showed that when clover was relied on for nitrogen supply, maximum yield was only obtained by using much more frequent watering, and more total water, than was needed when the nitrogen was supplied by fertilizer. Thus in a year with 'average' rainfall, maximum yield from a clover-grass ley was 86 cwt dry matter per acre; to get this needed 9 in. of irrigation water in the season. A similar yield was obtained by applying 34 lb N/acre before each cut and no water. When N was applied at 87 lb/acre per cut, the maximum yield of 121 cwt dry matter/acre was obtained with a total of only 3½ in. of water. Frequent watering of grassy swards dressed with nitrogen is wasteful. In wet areas, or where water is plentiful, the best can be made of clover. In dry areas, without irrigation, clover is much less useful and fixes less nitrogen; regularly high yields will only be obtained from N-fertilizer used on grassy swards. At Woburn, H. L. Penman[27] found that N-fertilizers and water were, to some extent, interchangeable for grass. Doubling the nitrogen dressing behaved as though it added ½ in. of extra soil water. The soil water deficits that checked growth were 1 in. when only 17 lb N/acre per cut were used but 1½ in. when 34 lb N were used. The increases in dry matter of a pure cocksfoot stand/acre/in. of transpiration given by 1 in. of water were:

N given lb/acre per cut	cwt of extra grass/acre
17	3·6
34	5·0

Arable crops

At Woburn, H. L. Penman[27] found that gains in yield of potatoes were 2·04 tons/acre/in. of potential transpiration when 112 lb N/acre was applied, and only 1·77 tons with half as much N. Extra N (45 lb as against 22½ lb N/acre) was more effective than irrigation in increasing barley yields, but spring wheat seemed to need the water before it could respond to extra N (90 lb as against 45 lb N/acre for this crop). Average gains for three-year periods with the higher rate of N were 2 cwt of extra barley grain from 1 in. of water and 7 cwt of extra wheat from 2½ in. of water. From American work, F. G. Viets[34] concluded that the consumption of water by field crops is 'about the same or is only slightly increased' by using fertilizers to grow much larger yields; thus doubling the yield by fertilizing does not double the water used. The value from a given amount of water can be increased where fertilizers can increase yields; also if fertilizers increase root development, more of the reserves of soil water can be used.

References

1. Adams, S. N. (1962). *J. Agric. Sci.*, **58,** 219.
2. Barnes, T. W., Clarke, R. T. and Dyke, G. V. (1963). *Rep. Rothamst. exp. Sta. for* 1962, pp. 183–6 and 193–7.
3. Barnes, T. W. and Clarke, R. T. (1963). *Rep. Rothamst. exp. Sta. for* 1962, pp. 1963–7.
4. Benzian, B. (1964). *Experiments on nutrition problems in Forest Nurseries*, Vols. I and II. Bulletin No. 37 Forestry Commission. (HMSO, London).
5. Birch, J. A., Devine, J. R., Holmes, M. R. J. and Whitear, J. D. (1967). *J. Agric. Sci.*, **69,** 13–24.
6. Boyd, D. A., Garner, H. V. and Haines, N. (1957). *J. Agric. Sci.*, **48,** 464–76.
7. Boyd, D. A. (1961). *Proc. Fertil. Soc.* No. 65.
8. Boyd, D. A. and others, (1962). *Rep. Rothamst. exp. Sta. for* 1961, pp. 173–80.
9. Boyd, D. A. and Dermott, W. (1964). *J. Agric. Sci.*, **63,** 249–63.
10. Bullen, E. R. and Lessells, W. J. (1957). *J. Agric. Sci.*, **49,** 319–28.
11. Carlson, C. W., Grunes, D. L., Haise, H. R. and Alessi, J. (1963). *Soil Sci. Soc. Amer. Proc.*, **27,** 319–23.
12. Clement, C. R. (1961). *J. Brit. Grassl. Soc.* **16,** 194–200.
13. Cooke, G. W. (1960). *J. Roy. Agric. Soc.* **121,** 184–7.
14. Cooke, G. W. (1963). *J. Roy. Agric. Soc.*, **124,** 144–9.
15. Cooke, G. W. (1964). *Proc. Fertil. Soc.* No. 80.
16. Dyke, G. V. (1963). *Rep. Rothamst. exp. Sta. for* 1962, pp. 183–6.
17. Fisher, R. A. (1924). *Phil. Trans. Roy. Soc. London B.* **213,** 89–142.
19. Gasser, J. K. R. (1961). *J. Sci. Fd. Agric.*, **12,** 562–73.
20. Holmes, J. C., Gill, W. D. and Rodger, J. A .B (1960). *J. Agric. Sci.*, **54,** 291–9.
21. Holmes, W. and MacLusky, D. S. (1954). *J. Agric. Sci.*, **45,** 129.
22. Jenkinson, D. S. (1964). *Rep. Rothamst. exp. Sta. for* 1963 p.55
23. Lessells, W. J. and Webber, J. (1965). *Expl. Husb.* 12, **62,** 74.
24. Low, A. J. and Armitage, E. R. (1959). *J. Agric. Sci.*, **52,** 256–62.
25. Low, A. J., Piper, F. J. and Roberts, P. (1963). *J. Agric. Sci.* **60,** 229–38.
26. Owers, A. C. (1960). *Experimental Husbandry Farms and Experimental Horticulture Stations Progress Report for* 1960, pp. 22–3.
27. Penman, H. L. (1962). *J. Agric. Sci.*, **58,** 343–79.

28. PIZER, N. H. and others (1961). *J. Agric. Sci.*, **56,** 197.
29. REITH, J. W. S. and INKSON, R. H. E. (1963). *J. Agric. Sci.*, **60,** 145.
30. SHAW, K. (1962). *J. Agric. Sci.* **58,** 145–51.
31. VAN DER PAAUW (1958). *Plant and Soil* **9,** 254–68.
32. VAN DER PAAUW, F. (1962). *Plant and Soil* **16,** 361–80.
33. WATSON, D. J. and PATTERSON, H. D. (1960). *Rep. Rothamst. exp. Sta. for* 1959, pp. 164–8.
34. VIETS, F. G. (1962). *Advances in Agronomy*. **14,** 223–64, Academic Press, New York.
35. WARREN, R. G. (1956). *Proc. Fertil. Soc.*, No. 37.
36. WIDDOWSON, F. V. and PENNY, A. (1963). *Rep. Rothamst. exp. Sta. for* 1962, 47–50.
37. WIDDOWSON, F. V., PENNY, A. and WILLIAMS, R. J. B. (1963). *J. Agric. Sci.* **61,** 397.
38. WIDDOWSON, F. V. *et. al* (1965). *J. Agric. Sci.* **65,** 45.

Discussion

Dr. R. C. Little asked *Dr. Cooke* if he thought that optimum nitrogen dressings would increase as much in the next twelve years as in the last. *Dr. Cooke* thought he should not reply but did comment that optimum dressings might rise; for example with greater use of irrigation and perhaps better disease control in crops.

Mr. L. J. Hooper asked *Dr. Cooke* to expand his views on water/nitrogen relationships, particularly with respect to summer rainfall. *Dr. Cooke* in his reply said that it was very difficult to disentangle the effects of rainfall on cropping system from the effects of rainfall on nitrogen response. However, it was very important that these effects should be disentangled and it was particularly important for grassland.

Mr. J. Webber said he noted with interest the data *Dr. Cooke* had presented on nitrogen residues from the fertilizer applied to the previous crop and asked *Dr. Cooke* whether he thought crops should be deliberately over-fertilized in order to get the benefit of the residues in the succeeding crop. *Dr. Cooke* replied that he thought there was a strong case for considering whether this was justified and said that more experiments on the extra yields obtained from nitrogen residues should be carried out.

Dr. J. Tinsley asked the speaker to explain the greater yield potential irrespective of nitrogen level in crops grown after trefoil compared with crops not grown after trefoil. *Dr. Cooke* replied that he had no explanation and could only assume that the effect was connected with slow nitrogen release from the trefoil.

Dr. D. B. Davies asked *Dr. Cooke* whether he thought the nitrogen residues from previous fertilizer use were in the form of nitrate distributed throughout the profile or whether the nitrate was converted into some organic form and then slowly released to the succeeding crop. *Dr. Cooke* replied that he did not have the necessary laboratory data to answer the question.

Mr. B. Wilkinson commented that he thought the effects of plant disease were not as simple as implied by the speaker; for example disease might affect the amount of nitrogen left over from a crop. *Dr. Cooke* replied that he thought very little was understood of the effects of disease on nitrogen response and quoted an experiment carried out by *Mr. Widdowson* of Rothamsted in which the yield of wheat was trebled by treatment of the soil with formalin. Nitrogen response was very small where formalin had not been applied but where it had been applied to control cereal root eelworm and take-all, large responses to nitrogen occurred. He thought considerable research was necessary to try to understand the effects of plant disease on nitrogen response.

Prediction of nitrogen requirements of arable crops following leys

By C. R. CLEMENT, B.Sc., Ph.D. and H. L. BACK, B.Sc.
Grassland Research Institute

It is not necessary to look far into the past to see that one of the main values of the ley in arable farming, was to build up the reserves of nitrogen in the soil through the agency of legumes and, probably to a lesser extent, through nutrient transfer by the grazing animal.

But there was, and still is, a very wide variety of swards and systems of grassland management and a correspondingly wide variation in the nitrogen status of the soil, and in the yield of crops which follow leys in rotation (Table 1).

Table 1

Experiments with winter wheat following a variety of 4-year leys
Extreme values of grain yield without nitrogen

	At Hurley (Expt. H9)	On Farms (Expt. HE26)
	Cwt/acre at 14 per cent moisture	
Highest yield	51	64
Lowest yield	17	5
S.D. of yield distribution	± 7·5	±10·7

This also means, as shown in Table 2, that the response of winter wheat after 3–4 year leys to fertilizer nitrogen cannot reasonably be reduced to any general term.

Table 2

Experiments with winter wheat following a variety of 4-year leys
Extreme values of responses of grain yield to 52 lb nitrogen/acre

	At Hurley (Expt. H9)	On Farms (Expt. HE26)
	Cwt/acre at 14 per cent moisture	
Highest response	14	20
Lowest response	−4	−18
S.D. of response distribution	± 4·1	± 6·5

It follows that there is no single meaningful value for the nitrogen requirement, for maximum production, of crops following leys.

There is, however, a need for guidance on the fertilizer requirements of crops after leys. The data shown in Table 3, taken from experiments carried out on commercial farms, suggests individual 'above average' farmers are somewhat uncertain as to the fertilizer to apply in this situation.

Table 3

The number of fields classified according to fertilizer nitrogen applied by the farmer to the crop and response to nitrogen obtained in trials in the same field. Wheat after leys Expt. HE26

Fertilizer nitrogen to field crop	Response in trial	
	Positive	Nil or Negative
Applied	41	23
Not applied	1	3

Even in the class where there was a response in the trial and the farmer used nitrogen (60 per cent of the fields), it remains questionable whether the correct amount of fertilizer was applied.

Are there any criteria which can usefully be employed to give greater precision to the nitrogen need of crops after leys?

The following tables and figures show some results from experiments at Hurley in which wheat yields after different leys have been compared in the same field for a number of years. Some preliminary results from an experiment (HE26) on a range of different farms, where winter wheat followed leys of 3–4 years duration are also presented. These extra-mural sites were found and selected with the assistance of Staff of the National Agricultural Advisory Service.

Information can be obtained from some aspects of the ley which can be observed directly in the field or extracted from farm records.

Legume content of the sward before ploughing. In trials at Hurley there has been a correlation between the legume content of the sward and the yield of wheat after leys. Figure 1,* for example, shows the relationship between the amount of lucerne in leys and subsequent wheat yield.

This relationship was also examined in the extra mural trial. Each sward was scored for legume content, using 50 throws of a 1½ in. square quadrat. There was, over all sites, a relationship between the values so obtained and wheat yield without fertilizer nitrogen. Although the regression significantly reduced yield variance, the residual variability is obviously too great to allow useful prediction of yield solely from legume content.

* In this diagram and in those which follow, the standard deviation about the regression is given as a measure of the residual variability in yield (or response) after removing that ascribed to the independent variables.

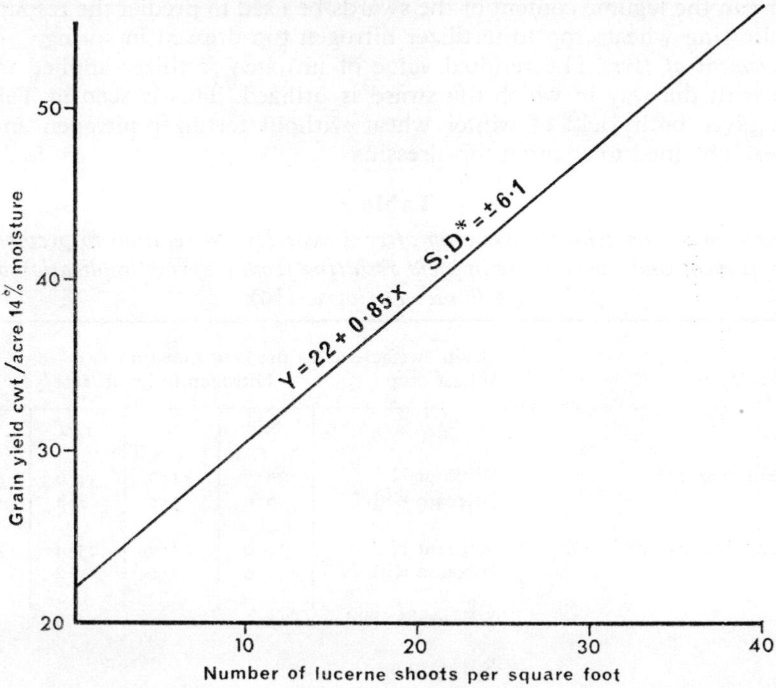

Fig. 1. Relationship between lucerne shoot numbers in leys before ploughing and the following wheat yields

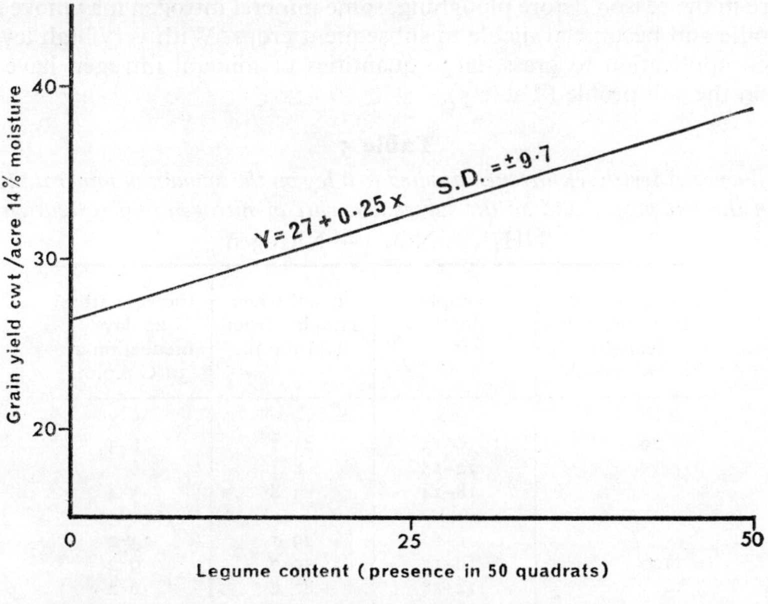

Fig. 2. Relationship between legume content of leys before ploughing and the following wheat yields. (HE26)

Nor can the legume content of the swards be used to predict the response of the following wheat crop to fertilizer nitrogen top-dressed in spring.

Management of leys. The residual value of nitrogen fertilizer applied to leys varies with the way in which the sward is utilized. This is seen in Table 4, which gives both yield of winter wheat without fertilizer nitrogen and the response obtained to a spring top-dressing.

Table 4

The yield of winter wheat (first crop after 4 year ley) in relation to previous ley treatment and the increase in yield resulting from a spring application of 52 lb nitrogen/acre (H9)

Previous ley use	Grain cwt/acre at 14 per cent moisture Wheat crop	Nitrogen to ley lb/acre/year Nil	70	140	280
Frequent grazing	Without N	32·7	34·0	34·6	40·2
	Increase with N	6·6	5·5	4·8	−1·0
Four cuts herbage removed	Without N	31·0	28·6	25·4	28·3
	Increase with N	7·6	7·1	8·2	6·3
	S.E. grain yield ±1·79				
	S.E. response ±1·49				

In general, nitrogen applied to grassland, only affects the following wheat crop when the herbage is grazed. However, there is some evidence in this table, and elsewhere, that under cut swards which receive more than 200 lb nitrogen per acre in the season before ploughing, some mineral nitrogen may move down the profile and become available to subsequent crops. With very high levels of nitrogen application to grass, large quantities of mineral nitrogen have been found in the soil profile (Table 5).

Table 5

The influence of fertilizer nitrogen applied to a ley on the amount of mineral nitrogen in the soil profile and on the subsequent rate of nitrogen mineralization

$NH_4^+ + NO_3^-$ — Nitrogen

Fertilizer to ley during previous year (lb N/acre/month)	Sample depth in.	In soil when sampled from field p.p.m.	Increase after 20 days incubation at 30°C p.p.m.
20	0–6	17·8	8·8
	6–12	24·2	6·3
	12–18	8·1	8·0
	18–24	4·8	5·4
100	0–6	49·1	13·2
	6–12	59·7	9·2
	12–18	82·2	6·2
	18–24	73·9	1·0

It may be worth noting that such levels of nitrogen use had little effect on the amount of nitrogen mineralized on incubation.

In the extra mural trials the individual farmers provided information on both the past manuring and utilization of the leys. As yet it has not been possible to distinguish any relationship between such data and either the yield of winter wheat after leys, or the level of response to nitrogen fertilizer; it is difficult to combine the sequence of levels and times of ferilizer application to give a meaningful numerical evaluation.

Winter rainfall. When leys are ploughed in summer or autumn, there is a rapid increase in nitrate-nitrogen in the surface soil. (A little less when the sward is killed with a herbicide and the land is not tilled.)

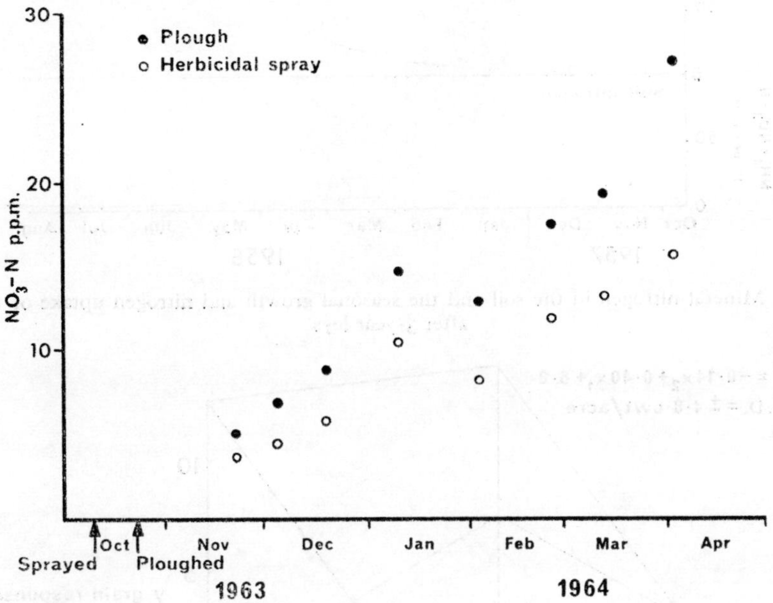

Fig. 3. Mineralization of nitrogen in an undisturbed and in a ploughed grassland soil

As shown in Figure 4, this mineralization of nitrogen can, during winter and early spring, exceed the rate of uptake by an autumn sown cereal and nitrate could be leached below the depth of subsequent recovery by the crop.

In common with results from trials with annual cropping there was a tendency, when wheat follows leys, for responses to be greater after wet winters. However, in the HE26 data winter rainfall and mineralizable nitrogen are correlated ($r = 0{\cdot}31$, $P < 0{\cdot}05$) and winter rainfall was only effective in reducing yield variablity when mineralizable nitrogen was held constant. Figure 5 shows the multiple regression of grain response on these two factors. The relationship accounts for about 20 per cent of the variation in response.

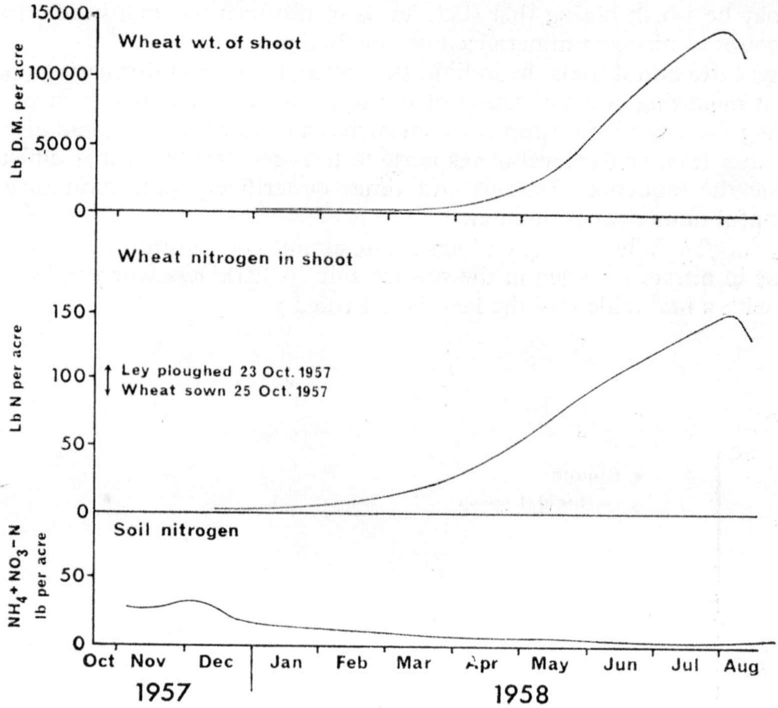

Fig. 4. Mineral nitrogen in the soil and the seasonal growth and nitrogen uptake of wheat after 3-year leys

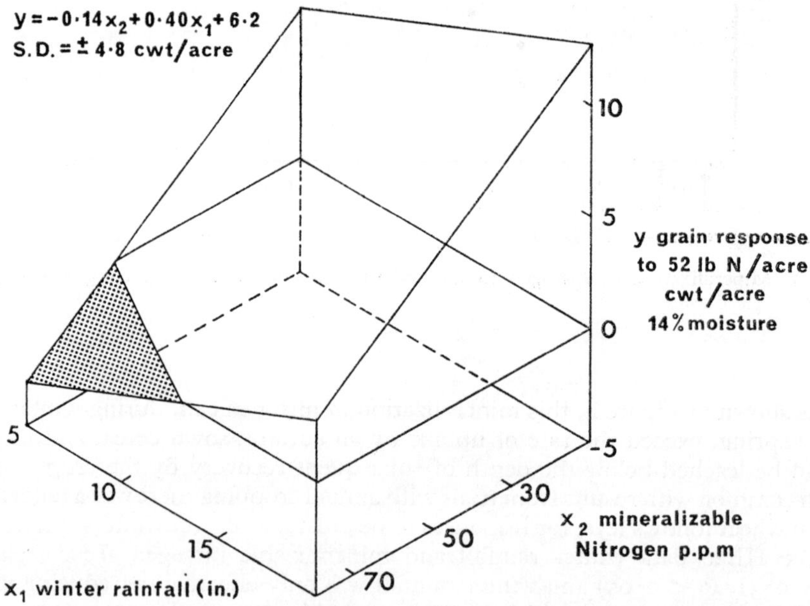

Fig. 5. Relationship between the response of winter wheat after leys to 52 lb nitrogen/acre, and soil mineralizable nitrogen and winter rainfall

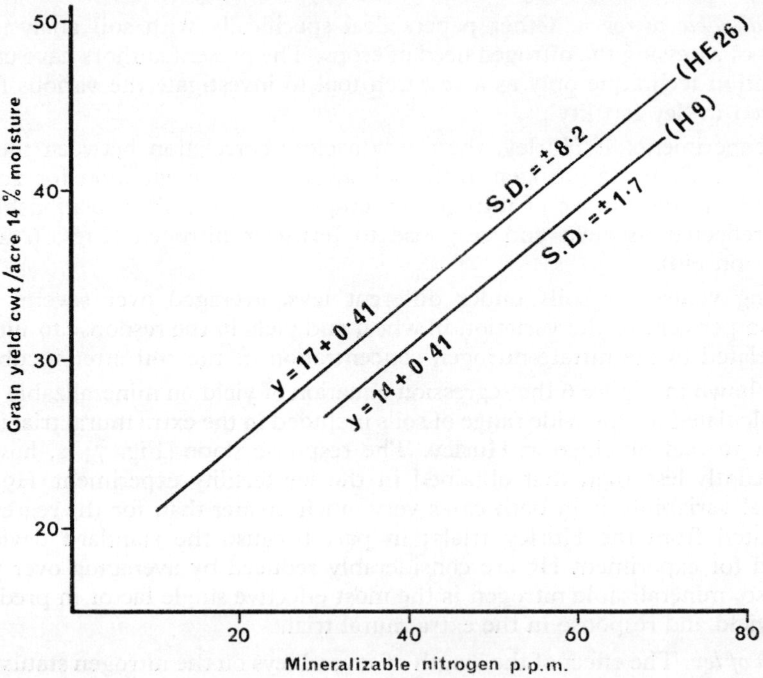

Fig. 6. Relationship between soil nitrogen status before ploughing leys and the following wheat yield

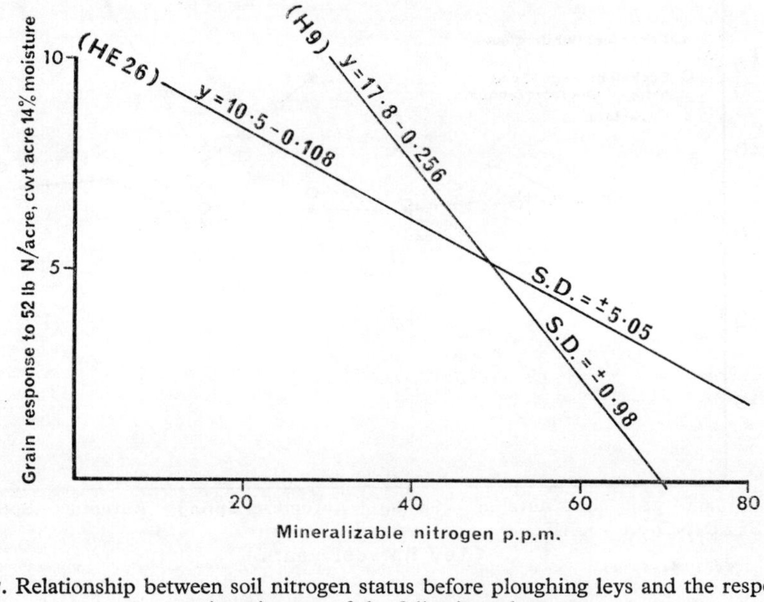

Fig. 7. Relationship between soil nitrogen status before ploughing leys and the response to spring nitrogen of the following wheat crop

Some effects of leys may be determined by soil analyses.

Mineralizable nitrogen. Other papers deal specifically with soil analysis as a means of assessing the nitrogen need of crops. The present authors have used an incubation technique only as a research tool to investigate the various factors involved in 'ley fertility'.

In experiments at Hurley, there was a close correlation between the concentration of nitrate-nitrogen in the soil after aerobic incubation for 20 days at 30°C and the uptake of nitrogen by crops after leys. This relationship has been reflected in yield and response to fertilizer nitrogen (Figs. 6 and 7, regression H9).

Using values for soils under different leys, averaged over several years, over 80 per cent of the variation in wheat and yield in the response to nitrogen was related to the nitrate-nitrogen concentration in the soil after incubation.

As shown in Figure 6 the regression equation of yield on mineralizable nitrogen calculated for the wide range of soils included in the extra mural trial is very similar to that obtained at Hurley. The response slope (Fig. 7) is, however, significantly less than that obtained in the ley fertility experiment H9. The residual variability is in both cases very much greater than for the regressions calculated from the Hurley trials; in part because the standard deviations quoted for experiment H9 are considerably reduced by averaging over years. Even so, mineralizable nitrogen is the most effective single factor in predicting both yield and response in the extra mural trial.

Length of ley. The effect of the length of several leys on the nitrogen status of an arable soil is shown in Figure 8.

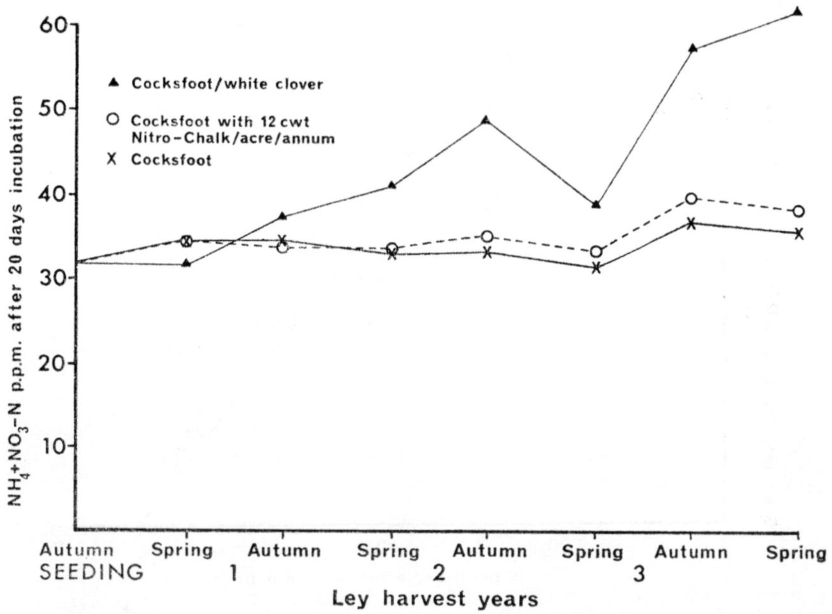

Fig. 8. Increase in the nitrogen status of arable soil under cut leys

It will be seen that under these swards, from which the herbage was cut and removed, only the grass/legume ley gave an increase in mineralizable nitrogen with time.

Other soil analyses. In the trials at Hurley many factors influencing yield and response of crop after leys were constant over the area of the experiment. With the trials on farms an attempt was made to measure several intrinsic soil characteristics which varied from site to site.

There were some significant relationships between these characteristics and yield; yield increased with clay content, silt content and with the water content of the soil at 75 cm suction. The relationships with silt and clay resulted mainly from their correlation with moisture holding capacity.

Response of wheat to nitrogen increased with stone content of the soil: possibly a relationship with soil depth.

It is evident that over a range of soils and localities many factors influence crop production after leys and the prediction of nitrogen requirement. A multiple regression analysis which included the following ten variables accounted for 60 per cent of the total variation in yield of wheat after leys on 63 sites included in the experiment HE26:

Legume content of the swards
Score for method of ley utilization
Rainfall October/March
Rainfall April/June
Rainfall July
Moisture capacity of soils at 75 cm suction
Mineralizable nitrogen
Clay, silt and stone content of the soils.

The same ten factors accounted for less than 30 per cent of the total variation in the response of the wheat crop to nitrogen.

Some of the remaining variation can be ascribed to experimental error and non-linear effects of the above factors, but there undoubtedly remain important factors which have not been taken into account in this analysis: for example, ineffective cultivations on several sites obviously led to serious competition between the wheat and grass weeds that could not be stated in numerical terms.

The analysis of the data from this experiment has not yet been completed.

In conclusion:

1. Different leys affect the response of wheat to nitrogen as much, or more than, soil types and localities. The management of the leys—their manuring, legume content and method of use—affects the soil nitrogen status and this, within a given soil type and climate, largely determines the yield and nitrogen response of the crops that follow. The soil and climatic factors which have been shown to affect nitrogen response after leys are moisture holding capacity, stoniness, and rainfall in the year of cropping.

2. If leys are ploughed when the soil is warm, there is an immediate release of nitrate which may exceed uptake by autumn sown crops. Wheat following leys showed a greater response to nitrogen applied in spring after a wet winter—presumably because some of the nitrogen in autumn was lost by leaching. Early ploughing may lead to greater loss.

3. Large amounts of nitrate have been found under grassland in early spring following high nitrogen inputs during the previous year. Retaining a grass cover during the winter months may reduce loss by leaching. This must be considered when fertilizing subsequent crops, grass or arable.

Acknowledgments

The authors wish to thank Dr. Wm. Davies and Mr. T. E. Williams for their help and advice in promoting this study and the National Agricultural Advisory Service and the many farmers who so kindly provided suitable field sites.

Discussion

Dr. C. W. R. Smith asked *Dr. Clement* if he had any direct information on the effect of time of ploughing of leys on the nitrogen response in the succeeding wheat crop. *Dr. Clement* said that he had no experimental data.

Mr. J. Webber commented that experiments at Experimental Husbandry Farms had shown no differences in the succeeding crop due to time of ploughing of leys.

Dr. D. B. Davies asked if there had been any attempt to classify, on a soil series basis, the soils at the experimental sites on privately owned farms. *Dr. Clement* replied that there had not and he thought this was a weakness. *Dr. Davies* said that in view of the amount of work done he was sure the N.A.A.S. would co-operate in classifying the soils if the speaker so desired. *Dr. Clement* agreed that this could be very worthwhile.

Mr. J. Brockman said that results similar to those given in Table 4 had been obtained at North Wyke but the experimenters there had concluded that the grazing animal was an important agent in the circulation of nitrogen. This appeared to be the reverse of *Dr. Clement's* observations. *Dr. Clement* replied that he had found under conditions of frequent grazing, only a small fraction of the nitrogen applied was actually utilized by the following crop.

Dr. G. W. Cooke asked how much potash had been applied to the herbage crop and how much to the wheat; did the potash application vary with the rate of nitrogen. What advice would the speaker give to farmers when growing wheat after leys. Had he any evidence on the effects of leys on succeeding crops other than cereals. *Dr. Clement* replied that basal potash dressings were applied to all treatments. In answer to the second question *Dr. Clement* said he thought that in giving advice to farmers the results obtained at the Grassland Research Institute could probably be adjusted with knowledge of local conditions, climate, clover content of the sward, etc. In answer to the third question *Dr. Clement* said that only cereals had been tried to date but experiments with potatoes would commence shortly. *Dr. Cooke* said his reason for asking the third question was that a crop such as potatoes or sugar beet had a main requirement for nitrogen later in the season than cereals and the effects of leys might be quite different. He quoted the effect found at Rothamsted of delayed release of nitrogen after ploughing a pure grass sward. *Dr. Clement* agreed but said that this delayed release of nitrogen had not been found at the Grassland Research Institute.

Determination of the nitrogen requirements of crops by analysis

(1) Laboratory methods of measuring soil nitrogen status and methods of correlating measurements with crop performance

By J. K. R. GASSER, B.Sc. (Nott.), Ph.D. (London), F.R.I.C.
Rothamsted Experimental Station

IN 1955 Harmsen and van Schreven comprehensively reviewed the mineralization of organic nitrogen in soil, including the use of mineral nitrogen content and mineralizable nitrogen to predict crop yields and responses to added fertilizers. During the past nine years results have been published of further investigations on the use of laboratory methods for measuring 'available'-N in soils, and various methods of correlating them with crop performance have been used. In addition to correlations of soil-N measurements with crop performance other results have been published of factors affecting mineralization of soil-N[11,12,20] and their effects on these correlations[14,15,20]. The present paper describes some of the methods used to measure 'available'-N, and discusses the ways of correlating laboratory measurements with crop performance. Some other factors affecting the correlations are also discussed.

Methods of measuring 'available'-N in soils

Fresh soils

Inorganic-N content: The nitrate-N and mineral-N (ammonium-N + nitrate-N) contents of fresh soils (nitrate-N_{fresh} and mineral-N_{fresh}) have been used as a measure of the 'available'-N[12,14,26].

Incubation: The nitrate-N and mineral-N formed on incubating fresh soils (Δ nitrate-N_{fresh} and Δ mineral-N_{fresh}) have also been used to measure 'available'-N[3,12,14].

Air-dry soils

Microbiological methods

Incubation: Many workers[4,8,9,10,16,17,23,24,26,27,29,30,34] have used the nitrate-N or mineral-N formed on incubating re-wetted air-dry soils (Δ nitrate-$N_{air\text{-}dry}$ or Δ mineral-$N_{air\text{-}dry}$), usually based on the method developed by workers at Iowa[11]. In this technique an inert material is mixed with the soil, which is maintained in a saturated atmosphere. Gasser[12] incubated soil alone at 50 per cent water-holding capacity.

Recently, Waring and Bremner[33] have suggested using the ammonium-N formed on waterlogged incubation of soil as a measure of the 'available'-N.

By growth of Aspergillus niger: The weight of mycelium formed by growing the fungus on soil covered with nutrient solution (except nitrogen) measured the 'available'-N in the soil[26].

Carbon dioxide produced from cellulose: Cornfield[6] suggested that the carbon dioxide produced on incubating soil treated with cellulose depends on the nitrogen supplied by the soil; therefore this production may be used as an index of the nitrogen status of soils.

Chemical methods

Permanganate oxidation: Various workers have used the ammonia released on alkaline permanganate oxidation of soil as a measure of 'available'-N[24,26,29,32].

Barium hydroxide extraction: Jenkinson[21,22] found that reducing polysaccharides extracted by 0·1 *N* barium hydroxide and measured as 'glucose' correlated with crop performance.

Ammonia formed on hydrolysis with sulphuric acid: Purvis and Leo[28] suggested the ammonia formed by hydrolysis on digesting soil with dilute sulphuric acid represented potentially 'available' nitrogen. Richard *et al*[29] used a similar procedure, but with more controlled conditions and modified to include nitrate.

Ammonia released by sodium hydroxide: Cornfield[5] suggested that ammonia released on treating soils with *N* sodium hydroxide may be related to their nitrogen supplying power.

Discussion

The most widely used measure of soil-N status is the nitrogen (as nitrate, or ammonium + nitrate) mineralized on incubating the re-wetted air-dry soils. This quantity appears to be a satisfactory estimate of the N supplying power of the soil, because many workers have found it correlates with crop performance. However, the estimation requires time to do because the soil samples must be air-dried, prepared, and stored for a period for some weeks before being re-wetted and incubated. The period of incubation varies from 7 to 28 days depending on the temperature and other experimental conditions. However, the method is suitable for classifying arable soils since samples may be taken at any time of the year; although values of mineralizable-N fluctuate with time of sampling, the changes are not great enough to alter the classification of the N status of the soils[8].

The nitrate or mineral-N contents of arable soils in spring has been found to be correlated with crop performance[12,14,15]. As nitrate is readily extracted from soils and estimated in the extract, the use of nitrate content appears useful as a guide to the N status of soils, and enables fertilizer dressings to be adjusted for the crop to be sown.

Chemical methods for estimating soil-N status are attractive because they allow more rigorous control of laboratory conditions and do not involve the time required for incubations. In the past the results with chemical methods were less good than with incubation techniques[24] but recently Jenkinson[21 22] found that the reducing sugars (measured as 'glucose') extracted from soils by 0·1 *N* barium hydroxide correlated well with N uptake by ryegrass grown in pots and with yields of barley grain from crops without fertilizer-N. However, Jenkinson reported that barium hydroxide extracted more reducing sugars than expected from soils of pH less than 6·0–6·5 and suggested that the question of the effects of soil pH on the extractable polysaccharides should be nvestigat ed further.

Methods of correlating laboratory measurements with crop performance

Direct correlation

In this method, yields of dry matter or N uptakes by the crops linearly correlated with soil measurements are usually expressed as the regression of yield or uptake on the soil parameter. The response to applied fertilizer-N is estimated by correlating increases in yields or uptakes at a given rate of application with soil measurement (in one case Pritchett *et al*[27] used a curvilinear instead of linear relation). If two or more rates of fertilizer-N are used the increase from a standard dressing may be interpolated from the parabola of best fit[12].

Percentage yield

The Mitscherlich equation relating yield to the supply of added nutritient has been widely used. Bullen and Lessells[2] describe it:

'The response curve used by Crowther and Yates was of the form:

$$y = y_0 + d(1 - 10^{-kx})$$

where y is the yield with a dressing of x cwt N per acre, y_0 the yield without fertilizer, d the limiting response (i.e., the response to a very large dressing) and k a constant which determines the rate of decrease in response.

The form of the curve implies that the response $(y - y_0)$ to a unit of fertilizer of amount p bears a constant ratio (10^{kp}) to the additional response to a second unit'.

Bray (1948) and Hanway and Dumenil (1955) have used a modified form of the equation to relate fertilized yield to unfertilized yield and a measurement of 'available' nutrient in the soil:

$$\log (A - A_0) = \log A - bx$$

where A = yield with nutrient, A_0 = yield without nutrient, b = constant and x = measured soil value. Considering the case where A is the maximum yield obtainable and expressing the results as a percentage of maximum yield Bray derived the form:

$$\log (100 - y) = \log 100 - bx$$

$$\text{where } y = \frac{\text{Unfertilized yield}}{\text{Maximum yield}} \times 100$$

This is a striaght line of slope b and intercept log 100, so that log (100 − y) and x may be linearly correlated and the usual test of significance applied to the coefficient, r.

Cook *et al*[4] also expressed their yields as a percentage of the maximum, but used a parabola to relate yields with soil test values.

Discussion

Advantages and disadvantages of linear and percentage correlations

Soil-N measurements are usually correlated either directly with crop yield without added-N or response to fertilizer-N or with the yield without added-N expressed as a percentage of the maximum yield.

The advantage of direct correlation is the ease of predicting crop performance from the regression line. Unfertilized yield increases and response to fertilizer-N decreases with increasing 'available' soil-N. The response to fertilizer-N may be more important than the unfertilized yield because it is the value of extra grain from the fertilizer-N applied that must interest the farmer. The response however is calculated either as the difference of the yields with and without fertilizer-N when it is determined less accurately than either, or as an interpolated point on the parabola of best fit, again subject to experimental error.

From the regression of response on soil-N measurement, the value may be calculated of the soil measurement at which zero response to fertilizer-N is expected, and this measurement used to calculate the unfertilized yield at zero response. The yield at zero response provides an estimate of the average maximum yield possible under the conditions of the experiments. This yield for 1958/59 is 41 cwt/acre calculated from Mineral-N_{fresh} and 44 cwt/acre from Δ mineral-$N_{air-dry}$.

Using percentage yields for correlating with soil measurements is valuable when an uncontrolled variable, such as weather, affects similarly yields of both unfertilized and fertilized crops. The disadvantage is that no direct numerical interpretation is possible; maximum yields must be known. These are difficult to determine on some soils and also the amount of fertilizer required to give maximum crop yields may vary between soils having similar unfertilized yields. Mathematically, this method excludes those results on which zero or negative responses are obtained to added fertilizer-N, because logarithms of zero and of negative numbers do not exist.

The value of the constant used in the modified Mitscherlich equation

The modified Mitscherlich equation derived by Bray[1] and subsequently used by many workers, assumes that the unfertilized yield is zero when the soil measurement is zero, but the experimental results suggest this is not so for nitrogen[12,14,36]. The linear correlations give an average unfertilized yield for barley grain of about 10 cwt/acre when measured soil-N is zero, and of 15 cwt/acre for winter wheat. Using a parabola to relate percentage yield to soil-N measurement, Cook *et al*[4] also found some yield was expected at zero soil measurement. The regression of percentage response on soil measurement also provides evidence that yields are not zero at zero soil measurement, because calculating the value of the constant in the regression equation,

$$\log(100 - y) = \text{constant} - bx$$

we have values of 1·932, 1·938, and 1·961, for regressions on Mineral-N_{fresh} in 1958, and Mineral-N_{fresh} and Δ mineral-$N_{air-dry}$ in 1959. (The correlation between percentage response and Δ mineral-$N_{air-dry}$ was not significant in 1958.) (From results of Gasser and Jephcott, 1964.) These correspond to unfertilized yields of 14·6, 13·3 and 8·6 per cent of the maximum; the average maximum yield in 1958 and 1959 was 37·0 cwt/acre, so the average yield at zero soil measurement is about 5 cwt/acre. As the intercepts are not accurately measured by extrapolation, the agreement between the two values must be regarded as satisfactory and the use of log 100 as the constant in the modified Mitscherlich equation may be regarded as incorrect. A lower value should be substituted. The change in the constant does not affect the shape of the curve (which is determined by 'b') but the smaller the constant the lower the soil value at which crops reach any percentage of the maximum yield. For example, in 1959 the calculated value of Mineral-N_{fresh} for 90 per cent yield was 22·5 ppm N using log 100 and 21·1 ppm N using the intercept value, log 86·7.

Effects of season on regressions of crop performance on soil-N measurements

Gasser and Jephcott[14] discussed the effects of season on the regressions of barley yield and response to fertilizer-N on soil-N measurements in two contrasting seasons:

'The late spring and summer (May-August) of 1958 were 'cool and wet' with temperature and sunshine below average and rainfall above average. Conversely 1959 was a 'warm dry' year. However, the season affects not only the growing conditions for the plants but also the incidence of pests and diseases.

Two seasonal effects that may affect the regressions of yield and response on soil-N measurements are (i) yields in individual years may increase at the same rate with increasing soil-N content but the mean values differ from year to year so that the regression lines are parallel (ii) yield or response may change in individual years at differing rates of soil-N content. Both effects show in our results.

The mean values for the two years calculated from the regression equation of yield on Δ mineral-$N_{air-dry}$ differed significantly and the difference between those of yield on Mineral-N_{fresh} approached significance, indicating that the regression of yield on soil-N measurement was described better by two parallel lines than by a single combined regression line. Therefore, although the rate of increase of yield with increasing soil-N was the same in each year, for a given soil-N value yield was greater in 1958 than in 1959. The slopes of the regressions of response on Mineral-N_{fresh} differed significantly in the two years and those of response on Δ mineral-$N_{air-dry}$ differed by a value approaching significance, indicating that the decrease in response to fertilizer-N with increasing soil-N status differed between 1958 and 1959. Therefore, a combined regression equation was less satisfactory than the separate relationships between response and soil-N status. The season also affected differently the degree of correlation between the various soil-N measurements and yield or response to fertilizer-N. mineral-N_{fresh} was better correlated with unfertilized yield and response to fertilizer-N in the wet (1958) than in the dry (1959) season. The reverse was true for Δ mineral-N_{fresh} and correlations with Δ mineral-$N_{air-dry}$ were little affected by differences between the two years. American workers[27] also found that regressions of responses of oats to fertilizer-N on soil-N measurements varied with season.'

Table 1 shows that soil-N measurements correlated better with crop performance in the individual years 1958 (13 soils) and 1959 (17 soils) than for the years 1957–63 (55 soils), suggesting that because of the seasonal effects on the correlations between soil measurements and crop performance, more information may be obtained by treating results from each year separately than by using combined results from a number of seasons.

Table 1

Effects of season on the value of coefficients of correlations between measurements of soil-N and yields of barley grain with and without fertilizer-N using two methods of correlating measurements with yield and response

Season and Soil Measurement	Method of correlation: Direct — Yield without Fertilizer-N	Method of correlation: Direct — Response to 0·5 cwt N/acre	Method of correlation: per cent Response
1958			
Mineral-N $_{fresh}$	0·83**	−0·91**	−0·74**
Δmineral-N $_{air\text{-}dry}$	0·72**	−0·61**	−0·53
1959			
Mineral-N $_{fresh}$	0·64**	−0·51*	−0·62**
Δmineral-N $_{air\text{-}dry}$	0·79**	−0·59*	−0·72**
1957–63			
Mineral-N $_{fresh}$	0·42**	−0·51**	−0·45**
Δmineral-N $_{air\text{-}dry}$	0·31*	−0·26	−0·06

* = Coefficient significant at 0·05 > P > 0·01 ** = P < 0·01

Errors in the use of samples of surface soil

Mineral nitrogen present in the sub-soil and the depth of ploughing are two factors which introduce errors into the estimation of the soil N status using samples taken only from the surface layer. Recent work has shown that on heavier soils considerable amounts of nitrogen may be carried over in the sub-soil from one year to the next[25,35]. This nitrogen may be either residues of fertilizer dressings or from mineralization of organic-N. Little is known of the amount of nitrogen in the sub-soil or of its value to crops. Eagle[9] estimated the effect of depth of ploughing on N status of soils by using the product nitrogen released on incubation (ppm N) × depth of ploughing (in.) as a measure of N status of the soils instead of the simple nitrogen release. He found the value of the correlation coefficient was increased from −0·66** to −0·76**, by estimating the effective cultivated volume of soil.

References

1. Bray, R. H. (1948). 'Diagnostic techniques for soils and crops.' p.53 American Potash Institute, Washington, D.C.
2. Bullen, E. R. and Lessells, W. J. (1957). *J. Agric. Sci.*, **49**, 319.
3. Clement, C. R. and Williams, T. E. (1962). *J. Soil Sci.*, **13**, 82.
4. Cook, F. D., Warder, F. G. and Doughty, J. L. (1957). *Canad. J. Soil Sci.*, **37**, 84.
5. Cornfield, A. H. (1960). *Nature, Lond.*, **187**, 260.
6. Cornfield, A. H. (1961). *J. Sci. Fd. Agric.*, **12**, 763.
7. Crowther, E. M. and Yates, F. (1941). *Emp. J. Exp. Agric.*, **9**, 77.
8. Eagle, D. J. (1961). *J. Sci. Fd. Agric.*, **12**, 712.
9. Eagle, D. J. (1963). *J. Sci. Fd. Agric.*, **14**, 391.
10. Eagle, D. J. and Matthews, B. C. (1958). *Canad. J. Soil Sci.*, **38**, 161.
11. Fitts, J. W., Bartholomew, W. V. and Heidel, H. (1955). *Proc. Soil Sci. Soc. Amer.*, **19**, 69.

12. GASSER, J. K. R. (1961). *J. Sci. Fd. Agric.*, **12**, 562.
13. GASSER, J. K. R. (1961). *J. Sci. Fd. Agric.*, **12**, 778.
14. GASSER, J. K. R. and JEPHCOTT, B. M. (1964). *J. Sci. Fd. Agric.*, **15**, 422.
15. GASSER, J. K. R. and WILLIAMS, R. J. B. (1963). *J. Sci. Fd. Agric.*, **14**, 269.
16. HAGIN, J. and HALEVY, J. (1961). *Israel J. Agric. Res.*, **11**, 81.
17. HAGIN, J. and RAVIKOVITCH, S. (1959). *Z. Pfl Ernahr, Dung.*, **84**, 110.
18. HANWAY, J. and DUMENIL, L. (1955). *Proc. Soil Sci. Soc. Amer.*, **19**, 77.
19. HARMSEN, G. W. and VAN SCHREVEN, D. A. (1955). *Advances in Agronomy* Vol. VII p. 300 Academic Press Inc. New York.
20. HARPSTEAD, M. I. and BRAGE, B. L. (1958). *Proc. Soil Sci. Soc. Amer.*, **22**, 326.
21. JENKINSON, D. S. (1963). *Rep. Rothamst. Exp. Sta.* for 1962, p. 61.
22. JENKINSON, D. S. (1964). *Rep. Rothamst. Exp. Sta.* for 1963, p. 55.
23. MACKAY, D. C. MACEACHERN, C. R. and BISHOP, R. F. (1959). *Canad. J. Soil Sci.*, **39**, 144.
24. MUNSON, R. D. and STANFORD, G. (1955). *Proc. Soil Sci. Soc. Amer.*, **19**, 464.
25. PAAUW, F. VAN DER (1963). *Plant and Soil* **19**, 324.
26. PETERSON, L. A., ATTOE, O. J. and OGDEN, W. B. (1960). *Proc. Soil Sci. Soc. Amer.*, **24**, 205.
27. PRITCHETT, W. L., BLACK, C. A. and NELSON, L. B. (1947). *Proc. Soil Sci. Soc. Amer.*, **12**, 327.
28. PURVIS, E. R. and LEO, M. W. M. (1961). *Agric. and Food Chem.*, **9**, 15.
29. RICHARD, T. A., ATTOE, O. J., MOSKAL, S. and TRUOG, E. (1960). *Trans.* 7th *Int. Congr. Soil Sci.*, **2**, 28.
30. SAUNDER, D. H., ELLIS, B. S. and HALL, A. (1957). *J. Soil Sci.*, **8**, 301.
31. STANFORD, G. and HANWAY, J. (1955). *Proc. Soil Sci. Soc. Amer.*, **19**, 74.
32. TRUOG, E., HULL, H. H. and SHILHATA, M. M. (1957). *New Soil test for available nitrogen.* Mineo Univ. of Wisconsin, Madison, W15.
33. WARING, S. A. and BREMNER, J. M. (1964). *Nature, Lond.*, **201**, 951.
34. WELCH, C. D. and BARTHOLOMEW, W. V. (1963). *Agron. J.* **55**, 441.
35. WIDDOWSON, F. V. and PENNY, A. (1963). *Rep. Rothamst. Exp. Sta.* for 1962, p. 50.
36. WILLIAMS, T..E., CLEMENT, C. R. and HEARD, A. J. (1960). *Proc. 8th Int. Grassland Congr.* p. 237.

Determination of the nitrogen requirements of crops by analysis

(II) Soil nitrogen release values and nitrogen responses in crops

By D. J. EAGLE, B.Sc., M.S.A.(Toronto)
*N.A.A.S. Wolverhampton**

Introduction

THE soil nitrogen test developed at Wolverhampton involves incubation of moist soil under controlled conditions and determination of the nitrogen mineralized. It is similar in principle to methods described by numerous other workers[2,7,8] and was adapted from a method developed in Canada[3] in which release of nitrate only was determined. In the present method release of nitrate plus ammonia is determined as it has been found necessary to estimate both in this country[4] because only calcareous soils release all mineralizable nitrogen as nitrate. The advantages of the method are that it is simpler and involves a shorter incubation period than other incubation techniques and it also has good reproducibility.

The method measures the capacity of the soil to release mineral nitrogen from organic matter. If the soil contains an appreciable amount of mineral nitrogen in the spring allowance must be made for it but the results of a recently completed national series of winter wheat experiments suggest that this is only likely to occur in low-rainfall areas.

Method

Ten gram samples of freshly dried and ground soil are placed between two layers of fine granular vermiculite in polythene weighing bottles (5·0 cm high and 2·5 cm diameter) with a small air hole in the base and in the cap. Sufficient distilled water is added to thoroughly wet the soil and the upper layer of vermiculite (5 ml). The excess water drains into the lower layer of vermiculite leaving a moisture content close to field capacity which is optimum for nitrogen release. The tube is then incubated at 35°C and high humidity for 7 days after which the ammonia and nitrate in the soil are extracted and determined. The quantity of ammonia and nitrate originally in the soil is determined on a separate sample and the total amount of nitrogen released is obtained by difference.

Air-drying and storage of samples

Air-drying of samples followed by re-wetting and incubation results in a greater release of nitrogen than incubation of the field moist soil. Fortunately the results from air-dried soils correlate better with crop response to nitrogen than the results from incubation of field-moist soils [2,7].

---

* Now stationed at Cambridge.

Storage of air-dry samples before incubation results in a further increase in the nitrogen release during incubation[4,6]. The increase is progressive for some weeks and the proportionate increase is far from constant for all soils. This means that only freshly dried soils can be incubated. Any storage of soils before incubation must be in the moist state under refrigeration. Soils should be incubated within two days of drying.

Correlation between yield response to N of annual field crops and nitrogen release values

The nitrogen release values (i.e., nitrate and ammonia) obtained have been correlated with response to nitrogen fertilizers in field experiments on several agricultural crops. Most of the experiments were on privately owned farms but a small number of the cereal experiments were at Rosemaund Experimental Husbandry Farm.

The modified Mitscherlich equation[1] was used as in previously published work.

This equation is:

$$\text{Log}\,(100 - y) = \text{Log}\,100 - b$$

$$y = \text{per cent of maximum yield} = \frac{\text{control yield} \times 100}{\text{maximum yield}}$$

$$c = \text{constant for any one crop}$$

$$b = \text{soil test value}$$

The correlation for winter wheat is shown in Fig. 1.

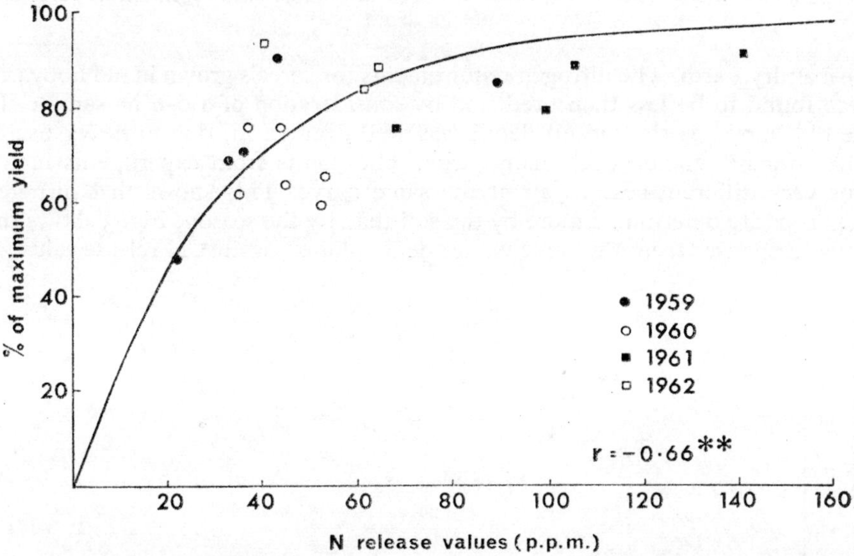

Fig. 1. Winter wheat: relationship between yield and nitrogen release values

The experiments were on soils ranging in texture from loamy sand to clay. Seven of the experiments followed potatoes, five cereal crops and six followed leys or permanent pasture. Some of the light soils were ploughed deeply, up to

15 in., while the medium and heavy soils were not ploughed so deeply. On deeply ploughed soils the organic matter containing the nitrogen reserves will be mixed with a greater amount of soil than if ploughing were less deep. Hence the nitrogen release values based on a standard weight of soil, irrespective of topsoil depth, will underestimate the nitrogen status of deeply ploughed soils and the product of nitrogen release value and ploughing depth should give a better estimate of soil nitrogen status[5]. As shown in Fig. 2 the correlation is considerably improved by taking the ploughing depth into account. This effect of topsoil depth has also been found at Rosemaund Experimental

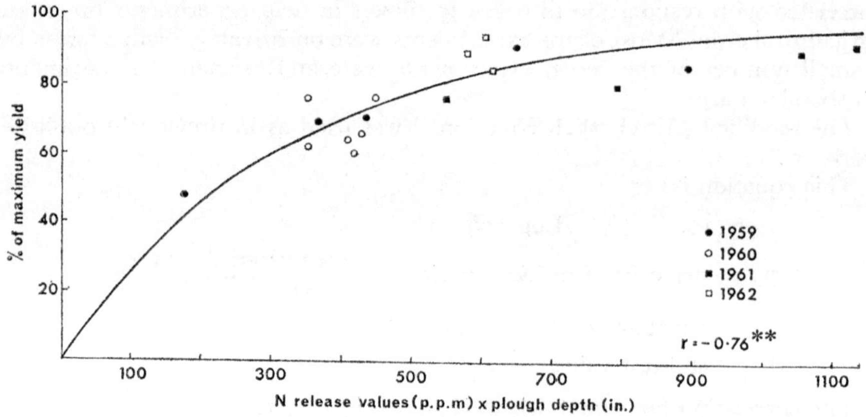

Fig. 2. Winter wheat: relationship between yield and product of nitrogen release value and plough depth

Husbandry Farm. The nitrogen requirements for cereals grown in old hopyards were found to be less than predicted by consideration of a 0–6 in. sample. In the old hopyards the topsoil depth was well over 12 in. due to heavy use of bulky organic manures for many years. The points from experiments in the four very different seasons all fit the same curve. This shows that nitrogen responses are determined more by the soil than by the season. Fig. 3 shows the actual responses from the same experiments plotted against N release value ×

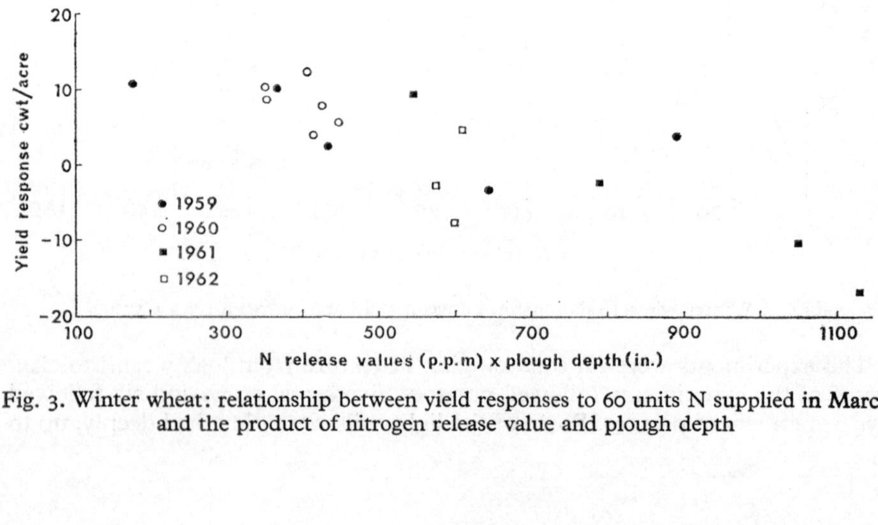

Fig. 3. Winter wheat: relationship between yield responses to 60 units N supplied in March and the product of nitrogen release value and plough depth

ploughing depth. The correlation is obvious and the points representing responsive and unresponsive sites are separated. This is included because there is no general agreement on the best method of correlation.

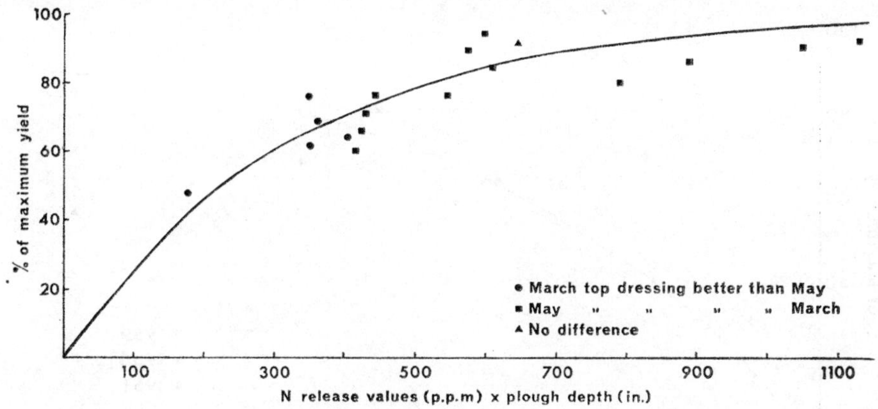

Fig. 4. Winter wheat: relationship between yield, product of nitrogen release value and plough depth and optimum time of top-dressing

Fig. 4 shows the relationship between optimum time of top-dressing winter wheat and soil nitrogen status. The soil test values give a rational explanation of why early top-dressing is optimal at some sites while late top-dressing is optimal at others. On deficient sites the early top-dressing gave the highest yields, presumably because adequate tillering was insured, while on high nitrogen sites the late top-dressing was more efficient. Severe lodging occurred at some of these high nitrogen sites when nitrogen was applied in March. It is well known that on heavy soils lodging can be troublesome in wet seasons but often not in dry seasons. In general, heavy soils are more likely to have high nitrogen reserves than light soils so a late top-dressing will usually be appropriate.

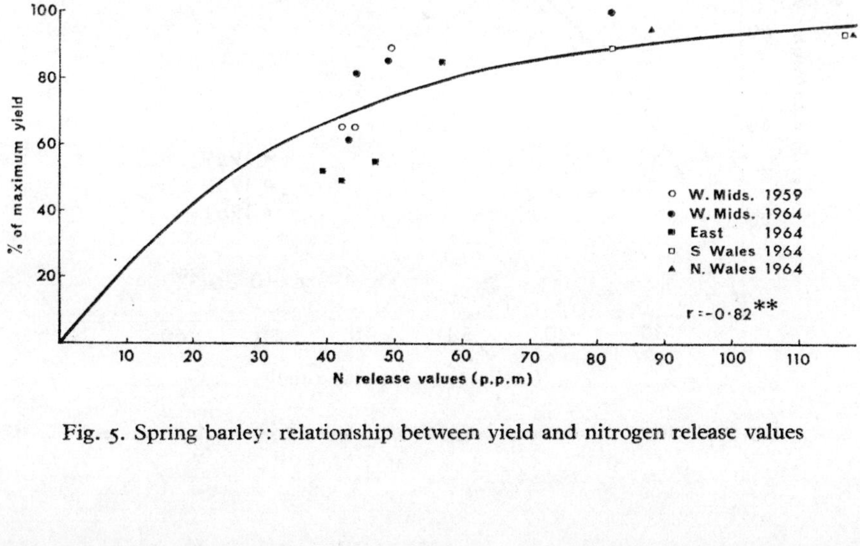

Fig. 5. Spring barley: relationship between yield and nitrogen release values

Fig. 5. Spring barley. Points from Eastern England and Wales all fit the same curve. This seems to suggest that the poorer response to nitrogen, usually found in high rainfall areas, is more associated with soil nitrogen status than with moisture supply.

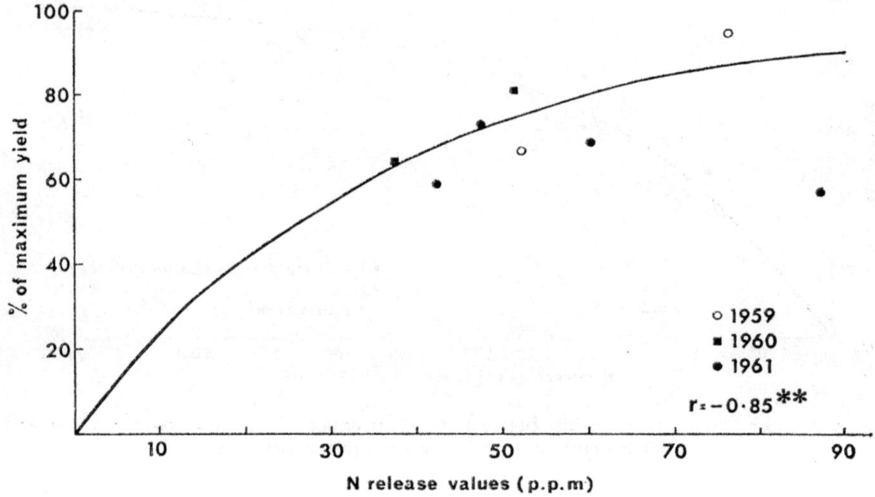

Fig. 6. Spring wheat: relationship between yield and nitrogen release values

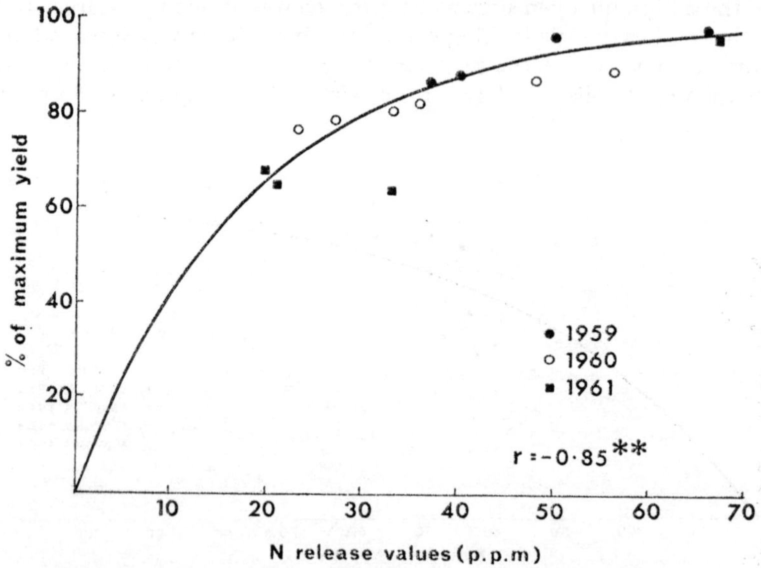

Fig. 7. Potatoes: relationship between yield and nitrogen release values

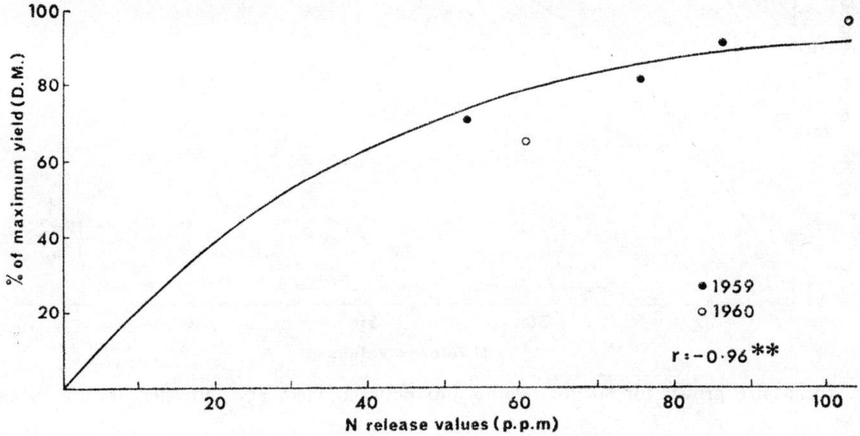

Fig. 8. Kale: relationship between yield and nitrogen release values

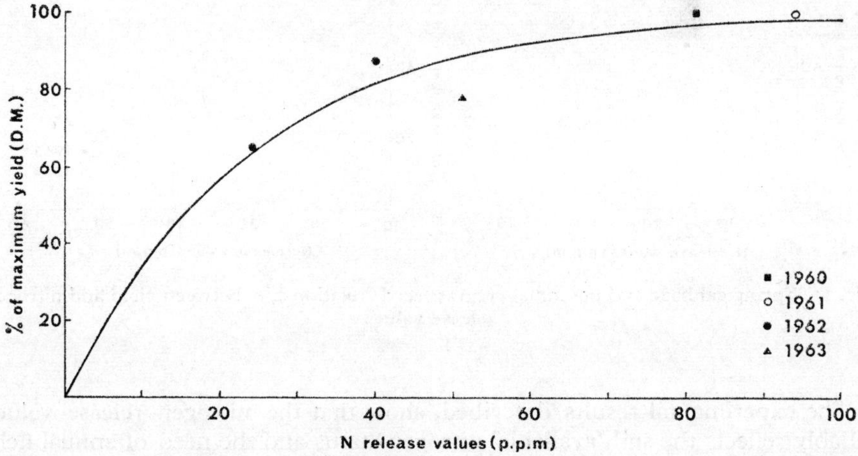

Fig. 9. Swedes: relationship between yield and nitrogen release values

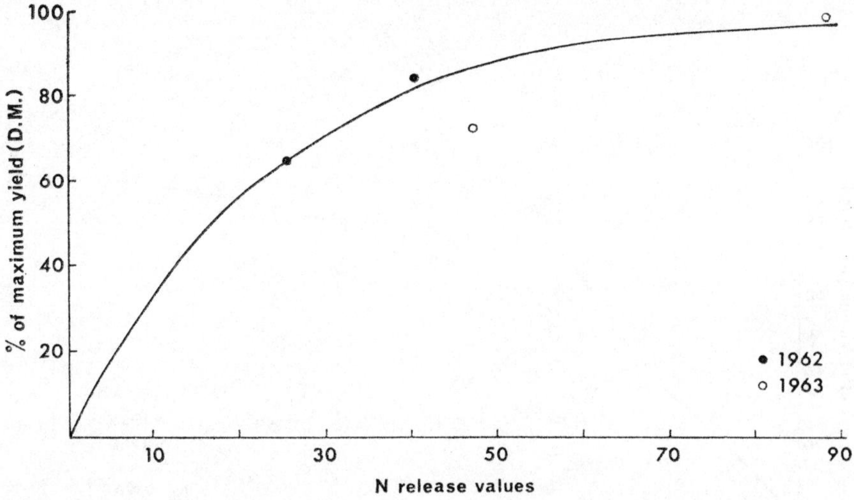

Fig. 10. Maize grown for silage: relationship between yield and nitrogen release values

Figures 6, 7, 8, 9 and 10 show the correlations for spring barley, spring wheat, potatoes, kale, swedes and maize, all of which are satisfactory. The only anomalous result was from spring wheat, grown on a soil with impeded drainage, which suffered from severe waterlogging during the exceptionally wet April of 1961. There have only been two experiments on spring cabbage and ryegrass seed but the results from these are shown in Fig. 11.

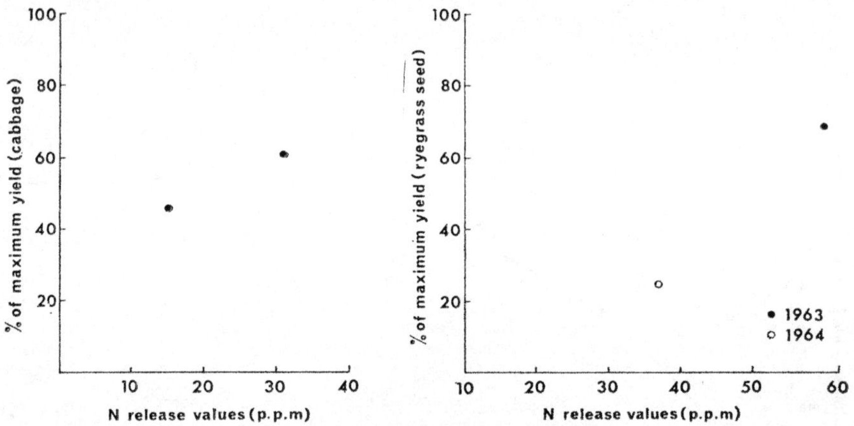

Fig. 11. Spring cabbage and perennial ryegrass seed: relationship between yield and nitrogen release values

The experimental results described, show that the nitrogen release values reliably reflects the soil 'available' nitrogen status and the need of annual field crops for nitrogen. Hence the method could be used for prediction of nitrogen requirements of crops.

Relation between nitrogen responses in blackcurrants and soil and leaf analyses

Two nitrogen experiments were started on established blackcurrant plantations in 1962. One site (Table I) was in Staffordshire on four year old Baldwin bushes. The other site was in Herefordshire on five year old Malvern Cross. The treatments, applied as ammonium nitrate/lime, were 0, 63, 126 and 189 units of N per acre.

Table 1

Soil	Year	Yield of fruit tons/acre		Per cent N in leaf DM (petiole removed)		N release value (control plots) p.p.m.	Remarks
		No N	189 units N	No N	189 units N		
Staffordshire							
Sandy loam	1962	3·70	4·37	3·03	3·37	24 V. low	No N applied in 1961
over clay loam	1963	3·09	4·34	2·64	3·20		
imp. drained	1964	3·14	4·38	2·40	2·98	18 V. low	Straw mulch
Herefordshire							
deep sandy	1962	4·47	4·45	3·12	3·62	23 V. low	Previous N use approx. 200 units per annum. Mulched with strawy poultry manure winter 1961–62
loam freely	1963	4·35	4·14	3·02	3·42		
drained	1964	1·34	2·14	2·77	3·25	17 V. low	

Nitrogen release values (Table 1) showed that both soils were deficient in nitrogen so a high nitrogen requirement night be expected. At the Staffordshire site, where no nitrogen had been applied in the year before the trial was laid down, a large yield response to nitrogen was obtained in the first year and larger responses in the two succeeding years. At the Herefordshire site, where previous nitrogen usage had been liberal and poultry manure applied as a mulch, no yield response occurred until the third year. Presumably the lack of response in the first two years was due to nitrogen from the mulch and residual nitrogen in the wood.

The samples for leaf analyses were taken in late June. The leaf nitrogen contents dropped progressively at both sites. The figures (Table 1) appear to suggest that no response to nitrogen will be obtained, in the year subsequent to sampling, when a leaf nitrogen content above 3·1 per cent is found while at the slightly lower level of 3·0 per cent N a large response to nitrogen may occur in the following year. This difference between sufficiency and deficiency is very small and the difference between the 1962 and 1963 leaf analyses from the control plots at the Herefordshire experiment was not statistically significant.

The soil nitrogen release values obviously cannot take account of residual nitrogen in the wood or nitrogen supplied from a surface mulch of bulky organic manure, so results from the soil in an established plantation could not give guidance as to the nitrogen requirement in any one season. However, a knowledge of the capacity of the soil to supply nitrogen might be of value in determining the over-all level of nitrogen manuring required.

Relation between nitrogen release values and nitrogen release from steam sterilized soils

It is well known that when a soil is steam sterilized the immediate release of nitrogen to a crop is greater than if the soil were not sterilized. Also nitrification is inhibited for a while so the level of ammonia builds up for some weeks until nitrification re-commences when there is a rapid fall in ammonia content and build up of nitrate. In an attempt to find whether nitrogen release values on unsteamed soil correlate with the amount of nitrogen released by steamed soils sixteen glasshouse border soils and potting loams were sampled before steaming and at intervals after steaming to determine the maximum ammonia level reached following steaming. The steaming was done by the grower following normal commercial practice. The soils were sampled weekly and analysed for nitrate and ammonia.

It was found that there was a high correlation between maximum ammonia level produced after steaming and nitrogen release value for all soils which contained 80 ppm or less of nitrate nitrogen before steaming. The other soils—all potting loams—contained from 150–420 ppm of nitrate nitrogen before steaming. These loams were all stored under cover so there was no leaching of nitrate by rainfall. Evidently the high nitrate content interfered with nitrogen release because some soils released no further nitrogen and in some there was a small decrease in nitrate content on incubation. The correlation for the low nitrate soils is shown in Fig. 12.

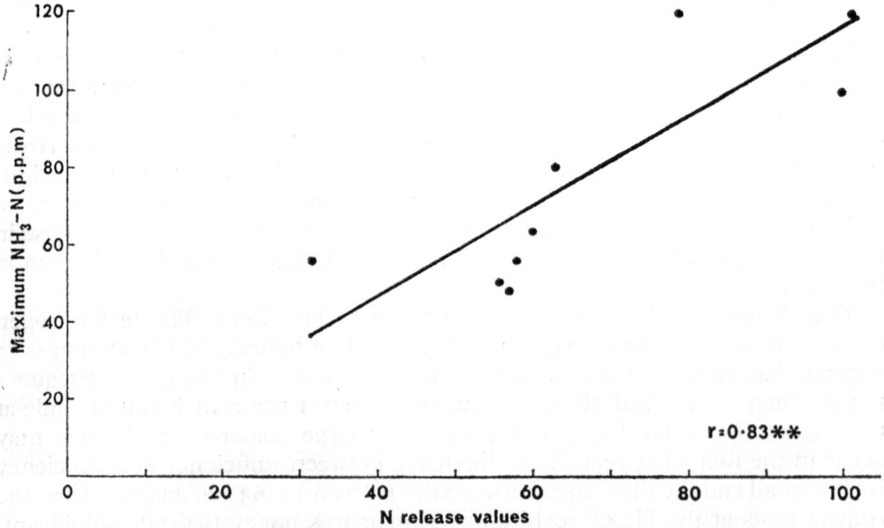

Fig. 12. Relationship between nitrogen release values and maximum ammonia accumulation after steaming

In a further study six soils of widely differing nitrogen status were prepared for incubation in duplicate incubation tubes. One of the duplicates was steamed in an autoclave for one hour at atmospheric pressure before incubation. The other duplicate was incubated in the usual way. The relation between the nitrogen release from steamed and unsteamed soil is shown in Fig. 13. It can be seen that the two sets of results are highly correlated. However, the results with and without steaming are practically identical which is surprising as sterilization is known to increase nitrogen release to crops. A possible explanation might be due to the fact the the soils were air-dried before sterilization. Air-drying also enhances the rate of nitrogen release.

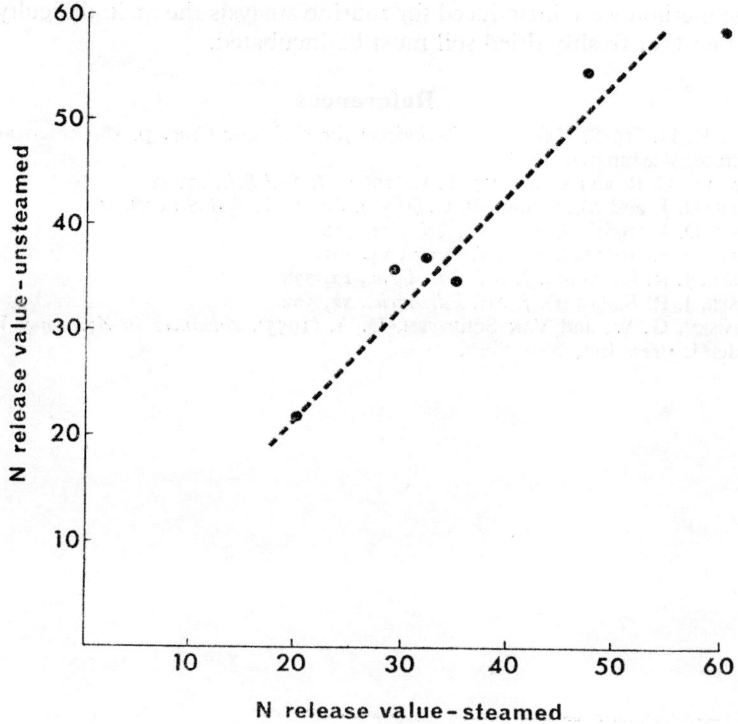

Fig. 13. Relationship between nitrogen release values of steamed and unsteamed soil

Application of incubation technique to routine analysis

As the incubation method described could be of value in both agriculture and horticulture the possibility of applying it as a routine method has to be considered. The main difficulty if the method were introduced would be the necessity for incubating freshly dried soil. All samples would have to be dispatched to the laboratory while still moist and then kept under refrigeration at high humidity until just before drying and subsequent incubation.

The placement of vermiculite, soil and water in the incubation tubes could be very rapidly done. For rapid removal of soil and vermiculite after incubation the base of the tube should be removable as well as the cap so that the contents could be easily pushed into the extracting vessel. Two determinations of ammonia (i.e., ammonia + reduced nitrate) are necessary—one after incubation and the other on the original soil. Apart from the seven day incubation period the determination of ammonia is the only time consuming operation as no determination of moisture holding capacity is necessary.

Summary

Soil nitrogen release values obtained by an incubation method were found to be highly correlated with the nitrogen response in several field crops.

Nitrogen release values on unsterilized soils were found to be highly correlated with the amount of ammonia released following steam sterilization, except when soils contained large amounts of nitrate before steaming.

If the method were introduced for routine analysis the main difficulty would be the fact that freshly dried soil must be incubated.

References

1. BRAY, R. H. (1948). *Diagnostic Techniques for Soils and Crops* p. 58. American Potash Institute, Washington, D.C.
2. CLEMENT, C. R. and WILLIAMS, T. E. (1962). *J. Soil Sci.*, **13**, 82.
3. EAGLE, D. J. and MATTHEWS, B. C. (1958). *Canad. J. Soil Sci.*, **38**, 161.
4. EAGLE, D. J. (1961). *J. Sci. Fd. Agric.*, **12**, 712.
5. EAGLE, D. J. (1963). *J. Sci. Fd. Agric.*, **14**, 391.
6. GASSER, J. R. K. (1961). *J. Sci. Fd. Agric.*, **12**, 778.
7. GASSER, J. R. K. (1961). *J. Sci. Fd. Agric.*, **12**, 562.
8. HARMSEN, G. W. and VAN SCHREVEN, D. A. (1955). *Advances in Agronomy VII*, 300. Academic Press Inc., New York.

Determination of the nitrogen requirements of crops by analysis

(III) Leaf analysis

By C. BOULD, M.Sc., Ph.D.
Long Ashton Research Station

EARLY workers in soil science and plant nutrition used plant analysis as a biological method for assessing soil fertility and they were concerned with the quantities of nutrients stored in the mature plant. Current workers, on the other hand, use leaf analysis, or chemical analysis of some other appropriate plant organ, as a guide to the nutritional status of the plant; first, to establish levels for nutrients below which plants show deficiency symptoms (confirmation of visual diagnosis), and secondly, to establish plant nutrient values associated with optimum growth, yield of some plant constituent such as sugar content, or with crop yield. Lagatu and Maume[15] were the first to adopt this new approach, to which they gave the name 'Diagnostic Foliare'.

The main object of this paper is to outline the general principles of leaf analysis and to discuss its application to the nitrogen nutrition of some crops.

General Principles

Salter and Ames[21], commenting on soil and plant analysis in relation to crop nutrition, stated that: 'All methods based upon a determination of the amount of a given nutrient present in the soil at a single sampling date and in such forms as to be extractable under controlled laboratory conditions, whether they be physiological or chemical in nature, are subject to the same limitations in that they disregard the differential feeding power of plants and the dynamic nature of the soil-plant relationship'. A chemical analysis of the particular crop grown on the particular soil under the particular local conditions to which the results will be applied has none of these limitations. The transformations which take place in the leaf constitute determinative processes that regulate growth and development. The leaf, therefore, may serve as an index tissue in the integration of all factors that influence the availability of soil nutrients and their uptake by the plant[26].

Macy[18], in an investigation of the relationship between leaf-N and the yield of oats, introduced the concept of 'critical nutrient percentages' in leaf dry matter. He held that for any given plant there exists a fixed 'critical percentage' for each nutrient; amounts in excess of this percentage represent 'luxury consumption', and amounts below represent a 'poverty adjustment range' that finally reaches a 'minimal percentage'. Macy found the minimal nitrogen percentage of oats at the milk stage to be 0·5 per cent N in dry matter, and the critical percentage as 1 per cent N in dry matter. Ulrich[28] advanced a slightly different interpretation and defined 'critical nutrient level' as that range of concentrations at which growth of the plant is restricted in comparison to that of plants at a higher nutrient level. Evidence shows that plants with widely different nutrient composition give similar yields so long as the nutrient concentrations are well above the critical levels.

Shear *et al.*[22] maintain that plant growth is a function of two variables of nutrition, intensity and balance, as they are reflected in the composition of the leaves when the plants are in the same stages of growth and development. At any level of nutritional intensity a multiplicity of ratios may exist between these elements. Maximum growth and yield only occur upon the coincidence of optimum intensity and balance.

Lundegärdh[17], working with oats, found a strong positive relation between S, or index, values (S = Spiegelwerten = mg atoms of element per 100 g of dry matter) and plant growth. He also showed that the response of oats to nitrogen depended on both the S_N and S_P index values. As the S_P value increased so the relative response to nitrogen also increased. This interaction between leaf nitrogen and leaf phosphorus has been noted by a number of workers[5,11].

Goodall and Gregory[12] maintain that growth is conditioned by two sets of factors: (1) the external factors such as light, temperature, water and nutrient supply and (2) internal factors, mainly nutritive but also hormonal. For each factor there is an optimum intensity level. Growth will be increased if the intensity is brought up to this level and decreased if it is raised further. These optima are not fixed but depend on all factors simultaneously; nevertheless, in theory one may postulate an optimal concatenation of factors at which development of the plant would be maximal.

It is clear, therefore, that the relationship between leaf nutrient concentration and crop yield is not a simple one because many factors other than nutrient supply may affect leaf composition[20]. Nevertheless it is possible, by factorial and response surface experiments, to establish the levels of nutrients at specific growth stages associated with deficiency, marginal response region and sufficiency[6] and so enable the advisory officer to make a more accurate diagnosis of the current nutritional status of a crop—an essential requirement before accurate advice can be given on fertilizer requirement.

Leaf sampling

The concentration of nutrients in leaves changes in a characteristic fashion as maturity and season advance. There is usually a rapid initial change in concentration during growth and enlargement of young leaves, followed by a plateau period of minimum change in the fully mature leaf[3], and finally a period of rapid change as senescence or dormancy approaches. It is most important, therefore, that leaf samples should be taken at specific physiological stages of development, not calendar dates, and from clearly defined positions on the shoot or plant. Sampling methods, for a wide range of crops, are described by Kenworthy[14] and Chapman[10], and sampling errors have been determined by Bould[4], Steyn[24] and Moon and Hymas[19] for a number of crops.

Nitrogen requirements of crops

Early workers expressed nutrient concentrations in terms of percentages in plant ash, based on the false hypothesis that the ash for a particular species was nearly constant, but it is usual now to express nutrient concentrations as percentage element in dry matter.

Tyner[27], working with corn (*Zea mays*), found that the concentration of nitrogen, in the sixth leaf from the base at 'silking' time, associated with optimum yield was about 3·1 per cent N in dry matter (2·9 per cent on air-dry matter). Above this level, response to fertilizer nitrogen was doubtful. Bennett *et al*[2], using the same sampling technique as Tyner, also found a definite relationship between yield of corn and the nitrogen and phosphorus content of the leaves. When the leaf-N reached 2·8 per cent in dry matter further increase did not appear to affect yield. Leaf analysis provided a clearer indication of the nutrient status of the plants than did analysis of the grain. Baird *et al.*[1], found the nitrogen levels associated with the economic and maximum crop yield of corn to be 2·4 and 2·5 per cent N in leaf dry matter respectively. Using multiple curvilinear regression analysis, Dumenil[11] found equal yields of corn associated with varying concentrations of leaf nitrogen and phosphorus, within certain limits, and he postulated that the critical level of N and P in the leaves is not a point, or narrow range of values, except at, or near, the maximum yield. Nutrient unbalance may occur when the leaf N or leaf P concentration extends beyond a certain limit in relation to the other one. (outside what he calls the 'ridge' lines, i.e., lines connecting the points on the successive isoquants having zero and infinite rates of substitution). Bould[5] found a similar relationship between leaf N and leaf P and the yield of raspberries. Slater and Goodall[23], working with lettuce, also found that the response to nitrogen depends on the level of leaf phosphorus. (see Table 1). Total leaf N was satisfactory for predicting response to nitrogen but nitrate-nitrogen concentration may enable a satisfactory forecast to be given at an earlier date. The analysis of different organs was too few to enable a satisfactory test of their relative value as diagnostic material to be made although they found no preference between older and younger leaves, or between either leaf and the whole tops. Ulrich[29] has made a comprehensive study of leaf analysis in relation to the nutrition and fertilization of sugar beets. He found the 'critical' nitrate-nitrogen concentration in the petioles of recently matured leaves of sugar beet to be approximately 1000 ppm of dry weight. (He defines the critical concentration as: 'that nutrient concentration which is just deficient for maximum growth'.)

Table 1

Nitrogen content of lettuce plants, grown in sand culture, corresponding to zero response to additional nitrogen supplies

(as percentage or p.p.m. dry matter)

29 days from sowing		44 days from sowing		
Phosphorus per cent	Nitrate nitrogen (p.p.m.)	Phosphorus per cent	Total nitrogen* per cent	Total nitrogen† per cent
0·2	1520	0·1	2·22	2·09
0·4	4290	0·2	2·68	2·65
0·6	7060	0·3	3·14	3·16
0·8	9830	0·4	3·59	3·54

* From linear expression

† From quadratic expression

(From Slater and Goodall, 1957)

The greatest use of leaf analysis in relation to nitrogen status and fertilizer requirement has been made with perennial crops. Leyton[16] studied needle composition in relation to the growth and nutrition of Japanese larch and Tamm[25] similarly for pine and other forest trees; Boynton and Compton[8], Heeney and Hill[13] and many others for apple; Bould[4] for soft fruits; Chapman[9] for citrus, to mention but a few of the workers using leaf analysis as a guide to the nutrition of perennial crops.

Heeney and Hill[13] found that a marked decrease in quality of apple fruits occurred as nitrogen levels in the foliage in mid-July increased above 2·15 per cent. Boynton and Compton[8] also found that when leaf N exceeded 2·0 per cent in dry matter fruit quality was reduced; below 1·85 per cent N in leaf dry matter yield may be considerably curtailed.

Bould[4] found that the optimum level of nitrogen in the recently-matured leaves of strawberry, at the fruit ripening stage, was between 2·6 and 3·0 per cent in dry matter, and for black currant the optimum level in leaves taken from the mid-third region of extension shoots just prior to fruit ripening was about 2·9 per cent in dry matter.

In general, a positive curvilinear relationship exists between leaf N at specific growth stages, crop yield and response to nitrogenous fertilizers. It has been shown that, in the absence of other limiting factors, the response to nitrogen is closely linked with the phosphorus status of the plant. Whilst it is possible, for a number of crops, to suggest tentative optimum levels of leaf N at specific growth stages for optimum crop performance, it is not yet possible to predict the amount of nitrogenous fertilizer required to raise the level of leaf N from a deficient state to a normal one.

References

1. Baird, B. L., Fitts, J. W. and Mason, D. D. (1962). *Proc. Amer. Soc. Soil Sci.*, **26**, 378.
2. Bennett, W. F., Stanford, G. and Dumenil, L. (1953). *Proc. Amer. Soc. Soil Sci.*, **17**, 252.
3. Bould, C. (1955). *J. Hort. Sci.* **30**, 188.
4. Bould, C. (1960). In 'Plant Analysis and Fertilizer Problems' (W. Reuther, ed.) Washington 6, D.C. *Amer. Inst. Biol. Sciences Pub. No.* 8.
5. Bould, C. (1963). *J. Sci. Fd. Agric.* **14**, 710.
6. Bould, C. (1964). *J. Sci. Fd. Agric.* **15**, 474.
7. Bould, C., Bradfield, E. G. and Clarke, G. M. (1960). *J. Sci. Fd. Agric.* **11**, 229.
8. Boynton, D. and Compton, O. C. (1945). *Soil Sci.* **59**, 339.
9. Chapman, H. D. (1960). In 'Plant Analysis and Fertilizer Problems' (W. Reuther, ed.). Washington 6, D.C. *Amer. Inst. Biol. Sciences Pub. No.* 8.
10. Chapman, H. D. (1964). *Foliar Sampling for Determining the Nutrient Status of World Crops. Part* 3, **16**, 36.
11. Dumenil, L. (1961). *Proc. Amer. Soc. Soil Sci.*, **25**, 295.
12. Goodall, D. W. and Gregory, F. G. (1947). 'Chemical Composition of Plants as an Index of their Nutritional Status'. (Aberystwyth: *Imp. Bur. Hort. and Plantation Crops*).
13. Heeney, H. B. and Hill, H. (1960). In 'Plant Analysis and Fertilizer Problems'. (W. Reuther, ed.) Washington 6, D.C. *Amer. Inst. Biol. Sciences*. Pub. No. 8.
14. Kenworthy, A. L. (1964). A Guide for Collecting Foliar Samples for Nutrient Element Analysis. Michigan, U.S.A. *Hortic. Dept. Michigan State University*.
15. Lagatu, H. and Maume, L. (1924). *Compt. rend. acad. sci.* **179**, 782.
16. Leyton, L. (1956). In 'Plant Analysis and Fertilizer Problems'. (P. Prevot, ed.) Paris: I.R.H.O.
17. Lundegärdh, H. (1951). *Leaf Analysis*. London: Hilger and Watts Ltd.
18. Macey, P. (1936). *Plant Physiol.* **11**, 749.
19. Moon, F. E. and Hymas, G. (1964). *J. Sci. Fd. Agric.* **15**, 201.

20. RICHARDSON, L. R. *et al.* (1954). *Influence of Environment on the Chemical Composition of Plants.* Southern Co-operative Series Bull. 36. Southern Exp. Stations and U.S. Dept. Agric.
21. SALTER, R. M. and AMES, J. W. (1928). *J. Amer. Soc. Agron.* **20,** 808.
22. SHEAR, C. B., CRANE, H. L. and MYERS, A. T. (1946). *Proc. Amer. Soc. Hort. Sci.* **47,** 239.
23. SLATER, W. G. and GOODALL, D. W. (1957). *Aust. J. Biol. Sci.* **10,** 253.
24. STEYN, W. J. A. (1960). In 'Plant Analysis and Fertilizer Problems'. (W. Reuther, ed.). Washington 6, D.C. *Amer. Inst. Biol. Sciences.* Pub. No. 8.
25. TAMM, C. O. (1956). In 'Plant Analysis and Fertilizer Problems'. (P. Prevot, ed.). Paris. I.R.H.O.
26. THOMAS, W. (1945). *Soil Sci.* **59,** 353.
27. TYNER, E. H. (1946). *Proc. Amer. Soc. Soil Sci.* **11,** 317.
28. ULRICH, A. (1948). In 'Diagnostic Techniques for Soils and Crops'. (Washington D.C. American Potash Institute).
29. ULRICH, A. (1960). In 'Plant Analysis and Fertilizer Problems'. (W. Reuther, ed.). Washington, D.C. *Amer. Inst. Biol. Sciences,* Pub. No. 8.

Discussion

Dr.K. Shaw opened the discussion by describing nine methods (four biological and five chemical) for determining soil nitrogen status. These methods had been investigated to a limited extent by the N.A.A.S. A bulk soil sample from each of thirty-three sites had been sub-sampled while field-fresh and distributed to several laboratories for analysis by different methods. (The methods are described in some detail in the appendix to the discussion.)

Eight of the nine methods gave results which correlated well between themselves and these eight also showed a similar pattern of relationship to the response of barley and winter wheat to added nitrogen; the exception was Method 6 (Appendix).

The same eight methods of analysis differentiated those soils which gave a low response to nitrogen in cereals, but only three biological methods, (Methods 1, 3 and 4—Appendix) showed promise for distinguishing between soils giving a moderate response and those giving a high response to nitrogen in cereals.

One method (Method 9—Appendix) had been studied in more detail over several years and had been correlated with nitrogen response in winter wheat, potatoes and kale. Generally, these correlations were poor and the method was only partially successful even when confined to trials on individual soil series.

These preliminary investigations were too limited to allow definite conclusions to be drawn, but it was considered that where methods appear to be of equal value, other factors such as ease of manipulation in the laboratory for subsequent adoption as a routine analytical method should be taken into account when deciding on which methods should be investigated more fully.

Mr. J. Webber said that Method 8 as quoted by *Dr. Shaw* had been tried on about 50 soils from experiments in Yorkshire and Lancashire and it appeared that the relationship between the analytical value and nitrogen response varied with soil type. In particular, results for Warp soils were quite different from those for other soils. *Dr. Shaw* suggested that if sufficient work were done on a range of soil types variability between soil types could be determined.

Dr. T. Z. Nowakowsky asked *Mr. Eagle* if, in his method, any allowance was made for fixation of ammonia by vermiculite. *Mr. Eagle* replied that the soil was incubated between two layers of vermiculite and there was no admixture, hence there should not be any significant amount of ammonia fixation.

Dr. G. W. Cooke suggested that the syndicates should discuss whether any laboratory procedure involving biological incubation under closely controlled conditions could imitate the process of nitrogen release in the field which was, in effect, another biological incubation process under very variable conditions of moisture and temperature. He asked *Dr. Gasser* and *Mr. Eagle* how the Mitscherlich equation assuming a logarithmic relationship could be used, if, as might well be the case, the response curve was parabolic. *Dr. Cooke* also suggested that it would be more important for advisory purposes to determine how often advice given to farmers, on a basis of classification of analytical results into categories of low, medium and high, would be correct than to calculate correlation coefficients.

Dr. Gasser said, in reply, that the parabolic form of equation for relating nitrogen response to analytical results had its place but like all mathematical representations it had its limitations.

Mr. Eagle said, in reply to *Dr. Cooke*, that the use of the Mitscherlich equation would not be affected if the yield response were parabolic, rather than logarithmic, because the yield without nitrogen is related to the maximum obtained. If excess nitrogen caused a reduction in yield the correlation would be unaffected.

Appendix to Discussion

Method 1 (Eagle, D. J., J. *Sci. Fd. and Agric.* 1961, **12,** 712)

The ammonia and nitrate produced on incubation of a moist soil is used as an index of the availability of nitrogen in the soil.

In this method the soil, immediately after air-drying and grinding, is incubated at a constant moisture content of about field capacity for seven days at 35°C. The constant moisture content of the soil is achieved by placing it between two layers of vermiculite in a narrow tube and adding sufficient water to moisten the soil and vermiculite. After incubation the soil vermiculite mixture is extracted with a $K_2SO_4-H_2SO_4$ mixture and the ammonia and nitrate in the extractant determined by the microdiffusion method of Bremner and Shaw[1]. The mineral nitrogen $(NH_4 + NO_3)-N$ produced on incubation is calculated by subtracting that present in the soil before incubation from that determined at the end of the experiment. Results are quoted as ppm $(NH_4 + NO_3)-N$ in oven dried soil.

Method 2

The basis of the nitrogen availability measurement in this technique is similar to that in Method 1.

The air-dried and sieved soil sample is moistened to 60 per cent of its water holding capacity (previously determined) and incubated in a stoppered bottle for 28 days at 25°C with regular aeration above soil level. After incubation a sample of the soil is extracted with $K_2SO_4-H_2SO_4$ mixture and ammonia and nitrate determined by the microdiffusion method of Bremner and Shaw[1]. Again the mineral nitrogen produced on incubation is calculated and expressed as in Method 1.

Method 3 (Waring, S. A. and Bremner, J. M., *Nature* 1964, **201,** 951)

The ammonia produced on incubation under anaerobic conditions is used as a measure of nitrogen availability in soil.

Air dried and sieved soil is incubated anaerobically under about 3 cm of water in a test tube for 16 days at 25°C. After incubation the whole contents of the tube are rinsed into a macro steam distillation apparatus with potassium chloride solution, magnesium oxide added and the ammonia distilled off, collected and titrated. The ammonia found before incubation is subtracted from the final ammonia value, thus giving that produced on incubation. Results are quoted as ppm of $NH_4 - N$ in air dried soil.

Method 4 (Waring, S. A., and Bremner, J. M., *Nature* 1964, **201,** 951)

This technique is identical with Method 3 except that the time and temperature of incubation is seven days at 30°C instead of 16 days at 25°C. The results are calculated and quoted as in Method 3.

Method 5 (Truog, E., *Com. J. Soil Sci.* 1959, **39,** 120)

The ammonia produced by alkaline permanganate hydrolysis of soil organic nitrogen is used as an index of the nitrogen supplying power of a soil.

Air dried and sieved soil is boiled for ten minutes with a mixture of potassium permanganate, sodium carbonate and water, the liberated ammonia being collected in water and determined by Nesslerization. Results are expressed as ppm NH_4-N in air dried soil.

Method 6 (Purvis, E. R. and Leo, M. W. M., *Ag. and Fd. Chem.* 1961, **9**, 15)

This method is based on the production of ammonia by hydrolysis of soil organic nitrogen with dilute H_2SO_4.

Air dried and sieved (1 mm) soil is digested to dryness with very dilute H_2SO_4 and then treated with a solution of trisodium phosphate. After filtering, ammonia is estimated by Nesslerization of the filtrate. The total NH_4-N present after hydrolysis is quoted as ppm NH_4-N in air dried soil.

Method 7 (Cornfield, A. H., *Nature* 1960, **187**, 260)

The ammonia produced by cold alkaline hydrolysis of the soil organic nitrogen with dilute NaOH is used as a measure of the nitrogen producing capacity of the soil.

Air dried and sieved soil is mixed with N–NaOH in the outer chamber of a microdiffusion unit and allowed to stand for 40 hours at 28°C. The ammonia liberated is collected in boric acid in the control chamber and determined by titration. Results are expressed as ppm NH_4-N in air dried soil.

Method 8 (Jenkinson, D. S., *Rept. Rothamsted Expt. Sta.* 1963, p.55)

The index of nitrogen availability determined in this method is the 'labile' fraction of soil organic matter, i.e., that which is soluble in 0·1N $Ba(OH)_2$.

Air dried and sieved soil is extracted by shaking with 0·1N barium hydroxide solution and the soluble carbohydrate content of the extract after filtering is determined by developing a colour with anthrone. The results are expressed as 'mg glucose' per 100 g air dried soil.

Method 9

This is a total soil nitrogen value as determined by a recommended Kjeldahl digestion procedure and the results are expressed as percentage N in air dried soil.

Reference

1. BREMNER, J. M. and SHAW, K. (1955). *J. Agric. Sci.* **46**, 420.

The role of nitrogen in grassland management

By J. O. GREEN,* M.Sc.
Grassland Research Institute

NITROGENOUS fertilizers are used on cash crops simply to increase their total yield. On grassland, they can be used to increase yield at particular times. The availability of nitrogen limits the growth of grass almost continuously in our climate. In order to realize the climatic potential for herbage production—four or five tons of dry matter per acre per year—the crop must have access to some 300 to 500 pounds of nitrogen per acre each year. This is at least three times the rate at which nitrogen is released from the majority of soils. The application of a relatively small quantity, say 50 lb per acre, of fertilizer nitrogen can be distributed in such a way as to raise yield only in those periods when growth is relatively slow. This will reduce the variation in rate of herbage production within the growing season. A great deal of emphasis has been placed, hitherto, upon the ability of nitrogenous fertilizers to change the pattern of forage production in this way and so to extend the growing season.

The extent to which forage production can be manipulated in this way calls for some discussion, but it will be useful first to examine the effect of nitrogen upon the total annual yield of herbage.

Response to fertilizer nitrogen

In the first place, the yield of a forage grass is much more closely related to the input of nitrogen than is that of a grain crop. The response of a pure grass sward to fertilizer nitrogen is usually linear up to a total annual input of 200, often 300, pounds of nitrogen per acre. The rate of response of a grass/clover mixture is also nearly constant over the same range of nitrogen levels, but it may be only half that of a pure grass sward (Fig. 1). In either case the magnitude of the response varies with the interval between harvests (Fig. 2).

If the interval between each nitrogen dressing and the following harvest is approximately six weeks, the response that may be expected from a grass sward is 20 to 25 lb dry matter per pound of nitrogen applied, and that from a well-balanced mixture of grass and clover about 10 lb dry matter per pound of nitrogen. In a grassy sward, and at present fertilizer prices, the cost per pound of dry matter will not change with level of nitrogen (except that in a short-term ley of Italian ryegrass, which carried a relatively high establishment cost, it may fall with increasing level of nitrogen). In the grass/clover mixture, the unit cost of forage will rise with increasing level of nitrogenous fertilizer. It will then be desirable to add nitrogen only if the increased cost of forage can be offset by a spreading of overhead costs and/or if the enhanced yield is associated with improved utilization. Both of these conditions are often fulfilled. If they are not, a limited input of nitrogen may still be used to increase profit per acre, so long as the extra forage replaces still more expensive foodstuffs. This will be considered under the heading 'Extension of the grazing season'.

* Due to author's illness the paper was read by D. W. Cowling, B.Sc., Grassland Research Institute.

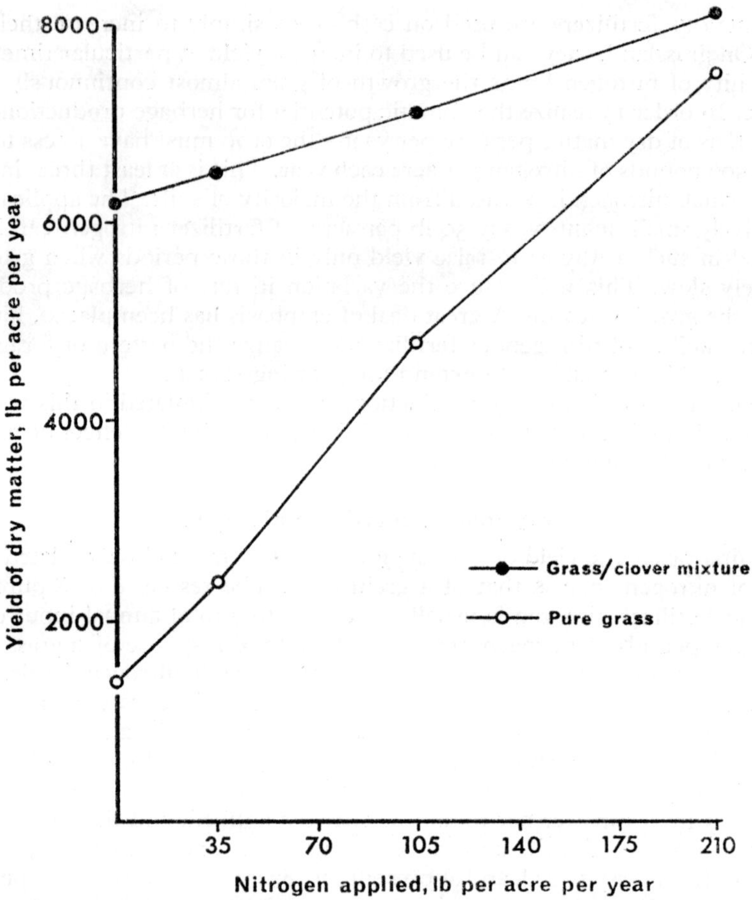

Fig. 1. Relationship between average annual yield of herbage and level of fertilizer nitrogen. 3-year leys of cocksfoot/white clover mixture and pure cocksfoot cut four times per year

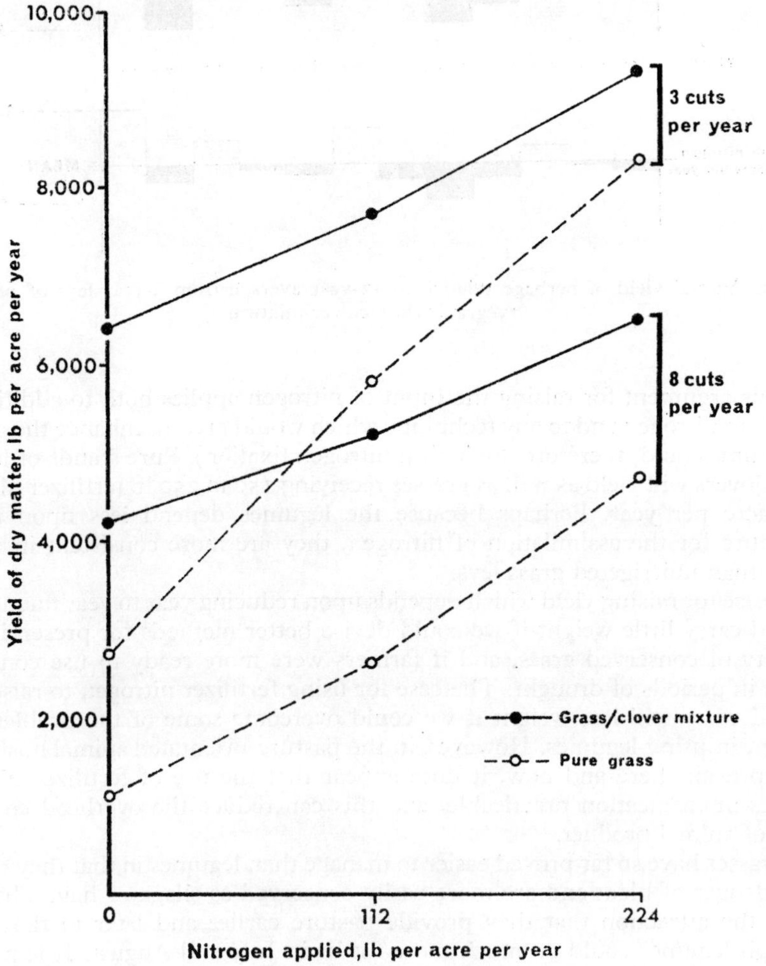

Fig. 2. Effect of frequency of harvesting upon the response of herbage to fertilizer nitrogen. Average of first and second harvest years in leys of perennial ryegrass/white clover mixture and pure perennial ryegrass

It is questionable whether a higher yield of forage in itself facilitates utilization in the short term. What may be more important, in present circumstances at least, is that variation in yield from year to year is somewhat reduced as the input of nitrogen is increased (Fig. 3). As the gross supply of forage becomes more predictable, so it becomes possible to intensify the stocking of grassland. That is to say, not only that the number of stock can be increased, but also that it can be higher in relation to the average annual yield of herbage.

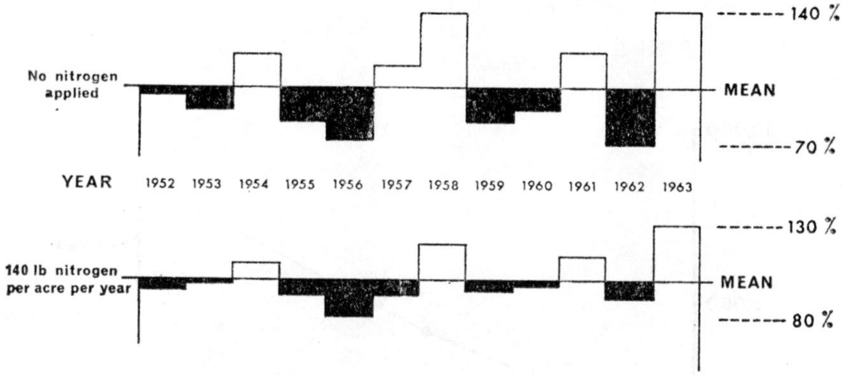

Fig. 3. Annual yield of herbage relative to 12-year average from 3-year leys of perennial ryegrass/white clover mixture

This argument for raising the input of nitrogen applies both to additions of fertilizer nitrogen and to any technique which would greatly enhance the growth of legumes (and, therefore, biological nitrogen fixation). Pure stands of lucerne and clovers will yield as well as grasses receiving 150 to 250 lb fertilizer nitrogen per acre per year. Perhaps because the legumes depend less upon topsoil moisture for the assimilation of nitrogen, they are more consistent in annual yield than unirrigated grass leys.

A case for raising yield which depends upon reducing year to year fluctuations would carry little weight if we could devise better methods for preserving the quality of conserved grass, and if farmers were more ready to use conserved grass in periods of drought. The case for using fertilizer nitrogen to raise yield would carry still less weight if we could overcome some of the problems involved in using legumes. However, in the pasture-orientated animal husbandry that prevails here and now, it does appear that the use of fertilizer nitrogen makes intensification practicable, and this can reduce the overhead costs per unit of animal product.

Grasses have so far proved easier to manage than legumes in that they present less danger of bloat and are more easily conserved as silage or hay. They also have the attraction that they provide pasture earlier and later in the year—though legumes could do much more for us in July and August. It is possible that the apparent benefit of nitrogen fertilizer to management is mainly a function of these attributes of the grasses, which respond so readily to it.

To return to the use of fertilizer nitrogen at different times of the year, it is important to consider, first the extension of the grazing season, and secondly the question of the summer depression in pasture production.

Extension of grazing season

Nitrogen can be applied for the specific purpose of accelerating growth early in the season, and it can be used to sustain better growth in autumn. The earlier the grass is used in spring, and the later the top-dressing is applied in autumn, the smaller is the return of dry matter per pound of nitrogen. One of the objects in top-dressing at these times is to increase the proportion of the total production that can be harvested directly by the stock, since grazing is the cheapest method of using grass. But care must be taken that the grass itself does not become too costly in relation to summer growth: for if early or late keep costs twice as much as summer grass it is no cheaper than silage.

The other object of extending the grazing season is to reap the benefit of any stimulus which fresh grass may confer upon growth or milk production. Although 'out-of-season' pasture may be rather costly, it can be profitable so long as it either improves the performance of individual animals (e.g., milk yield over the whole lactation), or at least replaces still more expensive balanced or high-protein concentrate feeds.

To be realistic in comparing the profitability of cows turned out one or two weeks earlier to grass with those that have to remain on conserved feeds, one should dismiss those herds that are being fed concentrates for every gallon—not to mention those that are getting some of their maintenance from concentrates—because, generally speaking, that sort of feeding reflects very inefficient conservation of the nutrients that can be produced cheaply from grass in summer. While such systems remain the common yardstick for measuring the value of early bite, and while the fertilizer-cost/milk-price ratio remains as favourable as it is (particularly in March/April), this seasonal use of nitrogen will no doubt continue.

One or two other points about fertilizer nitrogen applied for early bite should be considered. One is that early dressings are more effective in advancing the date of turn-out in an 'early' season than in a late spring. The date at which a reasonable quantity of grass could be offered from S.24 ryegrass in spring 1960 was brought forward, by a 60 lb dressing of nitrogen, from 12 April to 29 March (i.e., by 14 days): in spring 1963 a similar dressing brought it forward from 3 May by two or three days only.

Another point to note is that the harvesting of grass at an early stage of its development, and while it has a high nitrogen content, reduces the return of dry matter per pound of nitrogen applied. Also the removal of, say, 1000 lb dry matter per acre in early April can reduce the potential yield of a subsequent silage crop by about 2000 lb dry matter. Such data must be taken into account in the evaluation of early bite.

A further point is that nitrogen applied 'out of season' increases the total yield of forage, and, unless the extra grass actually replaces foodstuffs other than grass, it will raise the overall carrying capacity of the farm. If the stock number can then be increased proportionally, the increase will itself tend to shorten the grazing season again.

Finally, while the earliness of turning out to grass in any particular year is directly related to the quantity of nitrogen applied in February or March, so also is the residual effect that the early dressing has upon yield in May and June. This effect is so considerable that it must affect the year-round carrying capacity of the grass. A dressing of 60 lb nitrogen may give 29 cow-days as early bite, say 500 lb dry matter, but it will add probably another 500 lb dry matter to the yield of that grass by the time silage is cut. On a grass/clover mixture it will also depress clover production for several months. At this point one is bound to ask whether spring nitrogen alone does anything to improve the distributions of growth. It brings growth forward, but it also accentuates the peak in May and the shortage in July or August.

H

Summer pasture

Another tactical use of nitrogenous fertilizer, namely top-dressing to sustain growth after the spring flush has passed, has probably received less attention than it deserves.

Fertilizer nitrogen supplies a need in spring before soil nitrogen becomes available, but as the soil temperature rises there is a rapid increase in available soil nitrogen and, for a while the proportionate effect of fertilizer diminishes. It would appear, however, that quite early in the year the flow of nitrogen from organic reserves in the soil falls further and further below the capacity of grass to assimilate this nutrient. The deficiency of nitrogen in relation to the potential growth of grasses may become acute before the end of May. Whether the failure of soil nitrogen is due to exhaustion of readily mineralized material that has accumulated over winter, or whether it is due to increasing periods of moisture tension in the upper horizon, where the bulk of this material resides, is almost beside the point. The deficiency is all too apparent in a grass-dominant crop by early June, although the phenomenon is often mistaken for symptoms of drought.

In practice, there is a good chance that fertilizer nitrogen applied in late May and June will be effectively recovered by grass. Once in the plant, it will go a long way towards alleviating the slump in growth which occurs during July and/or August almost every year in the South-East and often in other parts of the country. Leys receiving top-dressings immediately after an early silage cut make rather more reliable summer pasture than leys dependent for nitrogen on the normal, small proportion of white clover; certainly more reliable than leys which have had only a spring top-dressing of nitrogen and consequently have even less clover in them by midsummer. At Hurley we have recorded some of the largest responses to fertilizer nitrogen in this period.

Compared with the predictability of grass production that can be achieved by irrigation, the effect of fertilizer nitrogen alone in midsummer may not be very impressive. It is nevertheless worth close study under farm conditions, since the stocking rate on many farms is largely governed by the amount of pasture available in the summer.

The great bulk of the nitrogenous fertilizer used on grassland is applied before the middle of May. If only because it is still proving difficult to conserve the nutriment so produced, a larger proportion of this nitrogen, and any additional quantity that may come to be used, should probably be directed towards production somewhat later in the season. If one has planned to exploit fertilizer nitrogen for aftermath pasture production, one has an added incentive to cut for silage or hay in good time, so that the next top-dressing can be applied while it has every chance of being effective.

Output per acre

Nitrogenous fertilizer may be used to extend the grazing season, but it may be better to regard it rather as a means of making better use of a long potential growing season to produce grass. Discontinuous top-dressing, to give a week or more two at pasture, is only a phase in the evolution of grassland fertilizer practice. The phenomenon of a near-linear response, for inputs of fertilizer nitrogen seldom attained yet in practice, suggests that the spotlight should be turned on to a higher total yield of grass over the whole growing season. This is particularly relevant where output per acre is the critical economic factor, and where an increase in stock numbers can be effected with minimal increase in overhead expenditure. The main conclusion I draw from the very interesting results published by I.C.I. Ltd., from their Henley Manor farm, is that the £7 per acre spent on nitrogenous fertilizer acted through its effect upon gross yield of grass, and through the higher stocking rate—and not on the length of the grazing season—to increase profit per acre from £15 to £39.

Discussion

Dr. G. W. Cooke commented on the value of legume nitrogen, compared with fertilizer nitrogen, for grassland. He said that grass could be grown when it was wanted with fertilizer nitrogen, but this was not the case when reliance was placed on secondhand nitrogen from legumes. He thought that farmers would increasingly use fertilizer nitrogen on their grassland. *Dr. Cooke* drew attention to the small amount of experimental work done with nitrogen on grassland and this was mainly centred around a few research stations. He agreed that lucerne was a high and consistent yielder of dry matter, as was red clover, and would fix about 200 lb of nitrogen per acre per year. However, in the drier parts of the country the amount of nitrogen harvested from grass-clover leys was only about 100 lb per acre per year. Faced with this situation he thought that farmers with high rents would have to use fertilizer nitrogen to produce grass economically. *Dr. Cooke* appealed for experimental information on the value of clover in leys and on the responsiveness of grass to nitrogen in different parts of the country.

Nitrogen and orchard management

By D. W. P. GREENHAM, M.A., Ph.D.
East Malling Research Station

The nitrogen requirements of the fruit tree

THROUGHOUT the growing season, nitrogen compounds are being utilized in the differentiation and growth of the various organs of the tree. Reserves can be accumulated, for instance in wood and bark over the winter, so that the seasonal distribution of nitrogen uptake by the tree does not necessarily reflect that of utilization. In practice it has been demonstrated[6] that nitrogen uptake by apple trees under cultivation starts in April and continues at least until July or August; it may well continue into the autumn for as long as root growth is active.

Much of the nitrogen taken up annually by the deciduous tree is returned to the soil, mainly in leaf-fall but also in abscissing blossoms and fruitlets and perhaps in leaching from the foliage by rain. The net annual removal from the orchard in a crop of apples will, in few cases, exceed 20 units of N per acre (a figure estimated for 1,000 bushels). However, from data for an American orchard cropping regularly at this high rate[1] it can be estimated that a further 20 units may be removed annually, partly in prunings and partly built into the main framework of the tree, giving a net annual nitrogen uptake of about 40 units per acre for a mature orchard cropping very heavily. The corresponding estimate for gross nitrogen uptake is about 90 units per acre. Satisfactory comparable data for English orchards have yet to be obtained, but it appears likely that lower figures may be expected in general; for example, an average crop of 300 bushels per acre contains only about six units of nitrogen.

The nitrogen supplied by a clean cultivated orchard soil

In clean cultivated orchards, apple trees often show no response to nitrogen applications for many years and, according to Chandler[3], in a considerable proportion of apple orchards in the U.S.A. the soil is able to supply as much nitrogen as the trees can use.

In a long-term manurial trial started in 1931 on dessert apples at East Malling, and maintained under a system of cultivation with autumn weeds, no consistent response to nitrogenous fertilizers (compared with no nitrogen) was seen for some time, but after 17 years the cumulative crop of all combinations of root-stock and variety showed a 15 per cent response. In 1953 half the plots were grassed down, but on the plots that are still cultivated the trees of Cox's Orange Pippin on M.XII (the only remaining rootstock-variety combination) on the −N and +N plots are still often indistinguishable in the field and show only slight differences in the concentration of nitrogen in their leaves (Table 1). The actual levels of leaf nitrogen might be interpreted as suggesting an adequate supply to the trees that had grown for 30 years without added nitrogen on soil by no means rich in organic matter or nitrogen reserves. In studying the seasonal trends of mineral nitrogen in the soil of these plots, G. C. White has shown that sufficient mineralization still occurs to give amounts of at least 50 units of mineral nitrogen per acre in each season, i.e., amounts of the same order as the estimated requirement of apple trees, perhaps explaining why the general nitrogen status of these trees appears to be satisfactory.

Table 1

Data from a long-term NPK trial on dessert apple trees (Cox's Orange Pippin/M.XII) at East Malling

		Cultivation		Sward (from 1953)	
		−N	+N	−N	+N
SOIL ANALYSIS 1958					
Organic carbon	0–6 in.	1·33	1·46	1·52	1·71
(per cent dry soil)	6–12 in.	0·82	0·84	0·97	0·93
Total nitrogen	0–6 in.	0·124	0·142	0·137	0·159
(per cent dry soil)	6–12 in.	0·095	0·096	0·105	0·102
LEAF NITROGEN	4.10.54	2·62	2·77	2·08	2·26
(per cent dry weight)	11.9.61	2·63	2·74	2·42	2·70
	20.8.62	2·81	2·92	2·59	2·94
	19.8.63	2·34	2·47	2·21	2·50
TOTAL CROP 1953–62 (lb per tree Cox's Orange Pippin/M.XII)		2602	3019	2232	3064

Soil: Free-drained fine sandy loam (Malling and Barming Series) overlying calcareous sandstone ('ragstone') of Hythe Beds.

Treatments started: 1931, on newly planted trees.

Management: Until 1953, cultivation with autumn weeds. In 1953, swards established on half the plots by sowing perennial ryegrass and wild white clover.

N treatment: Annual rate 30 units N per acre in 1931–40, 60 units 1941–52, 120 units since 1953; as sulphate of ammonia until 1958, then 'Nitro-chalk'.

Despite the similar appearance of trees in both treatments there have been large crop responses to applied nitrogen in recent years, amounting to a 25 per cent increase as the main effect of nitrogen over the 10 years 1953–62. The increase in weight of crop was due mainly to a greater number of fruits in the presence of nitrogen, though there was also an effect on fruit size. This result suggests that the nitrogenous fertilizer, in this case applied half in February and half in April, had an effect on the fruitfulness of the trees much greater than could be accounted for by the slight difference in general nitrogen status as indicated by leaf colour and nitrogen content in the summer. G. C. White's studies show the gradual build-up of mineral nitrogen in spring and summer, amounts remaining high until leaching of nitrate is brought about by late summer or autumn rain. During most of this period the cultivated soil may thus be able to meet the trees' nitrogen requirements, but there may be a critical specific requirement in relation to fruit set which cannot be met on the −N plots owing to insufficient build-up of mineral nitrogen at that stage in the season.

Nitrogen applications in cultivated orchards

Nitrogenous fertilizers are normally applied in the spring at rates of about 40–80 units of nitrogen per acre for dessert apples under cultivation and at higher rates for culinary apples. Rates of application are higher in the drier parts of the country, and on lighter soils, than in areas of high rainfall or in situations where more soil moisture is retained.

Such spring applications may act as useful supplements to the nitrogen made available by mineralization in the soil, and may both stimulate vegetative growth and improve fruit set. Foliar sprays of urea are also used at about blossom-time in many orchards with the object of improving fruit-set. During the summer the supply of mineral nitrogen in cultivated soil is probably adequate, as suggested above, and may be excessive for good fruit quality. However, in a wet year such as 1960, leaching of nitrate from the soil may occur in late summer, and under these circumstances a late summer or autumn application of fertilizer may be desirable. There is evidence from sand culture experiments both at Long Ashton[5] and East Malling (Delap, unpublished) that nitrogen given in late summer can stimulate fruit-bud initiation on new wood, while Williams[8], also at Long Ashton, has shown that nitrogen applied at this time can improve the effectiveness of the blossom by increasing the life of the embryo-sacs. The application of relatively concentrated (4 per cent) urea sprays in the autumn just before leaf-fall is also being tried in this country following reports of increased fruit-set from this treatment in Norway[7]. There is no evidence so far that these late urea sprays are effective in Britain and responses to nitrogen applied late in the year may perhaps be shown only with trees of generally low nitrogen status.

The effects of grassing down

Grass swards compete with fruit trees for water and for nitrogen, the severity of the competition depending on the relative vigour of sward and tree. With suitable management, i.e., with grasses that are not too vigorous for the tree, mown frequently, and with extra nitrogen added, the check to the growth and cropping of the tree on grassing-down may be minimized. Under such circumstances, the period of check may last only for two to four seasons.

The grass-clover swards sown in 1953 on half the plots in the East Malling NPK trial (Table 1) reduced the nitrogen status of the trees in the first few years. On the plots receiving no nitrogen, this reduction in nitrogen status was severe at first and was still evident after ten years; compared with the plots still being cultivated, the total weight of crop over the ten years was seriously reduced. Where nitrogen was applied (at 120 units per acre per annum), however, the swards ceased to depress the nitrogen status of the trees after four years and there was no loss in total crop (Table 1).

That the nitrogen status of trees in plots receiving nitrogen should be unaffected, after the first few years, by the presence of a grass sward is somewhat surprising in view of the reputation of grass for efficiency in uptake of mineral nitrogen; in this trial the amount of mineral nitrogen in the soil of the sward plots remained low. Excavations of root systems of apple trees at East Malling[4] have shown that under grass there is a much greater fibrous root development than under cultivation, suggesting that trees in grass may maintain their nitrogen status by more thorough exploitation of the soil, compensating for the reduced amounts of mineral nitrogen in the soil.

It appears that the advantages in fruit colour and storage quality associated with grassing down are largely due to the restriction of the nitrogen supply to the tree, for in the East Malling trial these advantages tended to disappear as the nitrogen status of the trees returned to the level of those in the cultivated plots. This suggests that, although the rate of nitrogen application must be increased on grassing down (by about 40–60 units per acre), and must remain high for one or two further years, it may profitably be reduced thereafter to its former level.

Orchard swards may be regarded as a means of regulating the tree's seasonal supply of nitrogen to the best advantage. Grass species differ in the severity of the competition they offer, and clover, in conditions where it persists, may actually increase the nitrogen supply to the tree[2]. Frequency of cutting also markedly affects the competition for water and nitrogen.

Urea sprays are a potentially useful means of boosting the tree's nitrogen supply at critical stages of development, such as the time of setting of the blossoms, in combination with sward management to achieve good fruit quality by restricting the general nitrogen supply.

The effects of mulching

Overall mulches of straw have been found at East Malling to stimulate the growth of fruit trees, probably as a result of the conservation of soil moisture and of the suitable conditions created for root growth in the topsoil. When such mulches were allowed to remain for several years, the nitrogen status of the trees was increased; only in the first season was there any evidence of a restriction in the nitrogen supply to the tree.

Overall mulches are not commercially practicable for tree fruits, but individual tree-mulches are of value in reducing the competitive effect of a sward. Herbicides, with or without such mulches, may allow an orchard to be grassed down at an earlier age than would otherwise have been feasible; the establishment of a sward at planting, or even planting into established grass, are practicable even in dry areas if the trees are protected from competition in this way for a few years.

The effects of irrigation

In those situations where it is economic to irrigate orchards, the growth of the trees is increased by the irrigation, and if the trees should become over-vigorous there may be a need to reduce the nitrogen applications. Where both soil moisture and nitrogen are limiting, however, the responses to both water and nitrogen applications may be additive, and it may be necessary to apply extra nitrogen if the increased cropping potential represented by the extra growth is to be realized. There are indications, for instance, that irrigation has a greater effect on fruit size if nitrogen is applied.

References

1. BATJER, L. P., ROGERS, B. L. and THOMPSON, A. H. (1952). *Proc. Amer. Soc. Hort. Sci.*, **60**, 1.
2. BOULD, C. and JARRETT, R. M. (1962). *J. Hort. Sci.*, **37**, 58.
3. CHANDLER, W. H. (1957). *Deciduous Orchards*, 3rd edn. Henry Kimpton, London.
4. COKER, E. G. (1959). *J. Hort. Sci.*, **34**, 111.
5. HILL-COTTINGHAM, D. G. (1963). *J. Hort. Sci.*, **38**, 242.
6. MASON, A. C. and WHITEFIELD, A. B. (1960). *J. Hort. Sci.*, **35**, 34.
7. OLAND, K. (1963). *Nature*, **198**, 1282.
8. WILLIAMS, R. R. (1963). *J. Hort. Sci.*, **38**, 52.

Discussion

Dr. T. H. Rose referred to the leaf analysis data in Table 1 which gave leaf nitrogen figures for Cox's Orange Pippin ranging from 2·34 to 2·81 in various years and asked *Dr. Greenham* if a response was obtained in every year. *Dr. Greenham* replied that yield responses to nitrogen, which did not vary greatly from year to year, were obtained in each year. *Dr. Rose* asked how leaf analyses could be interpreted in view of this large variation between years. *Dr. Greenham* said that the variability was typical of what was found in practice. He thought the yield was determined by nitrogen supply at some critical stage of growth, and was not necessarily reflected by the leaf analyses on samples taken in August.

Dr. C. Bould quoting from data by Bollard in New Zealand, showed seasonal variations in nitrogen content of the xylem sap of fruit trees with a peak at blossoming time. He thought this stage was very critical as it influences the longevity of the embryo sac, therefore the amount of blossom which sets and hence the amount of fruit. Thus although leaf analysis gives a general guide to the nitrogen status of the tree, there were other factors to take into account.

Mr. R. Thomas asked about pruning of the fruit trees in the experiment providing the data in Table 1 as he thought pruning would affect leaf nitrogen content. *Dr. Greenham* agreed that pruning would be one of the many factors affecting leaf nitrogen content. The trees in the experiment were pruned lightly on all treatments.

Nitrogen and glasshouse crops

By G. W. WINSOR, B.Sc., Ph.D., A.R.I.C., D.I.C.
Glasshouse Crops Research Institute

MANY of the problems associated with the use of nitrogen under glass arise from the properties of nitrates, this being the form either in which nitrogen is applied or to which it is generally transformed in the soil. Nitrates are highly soluble salts, and the nitrate ion is not appreciably adsorbed; thus nitrates are readily leached from the soil, yet in excess they constitute a major factor in man-made salinity problems. The transformation of nitrogen between the organic and inorganic forms is also of importance in glasshouse cropping, since partial sterilization by steaming, which favours mineralization of nitrogen, is widely practised.

Any attempt to review the subject as a whole would lead to repetition of much that is already well-known to soil chemists and advisers. Two aspects of the subject have therefore been selected in which much remains to be studied and correlated. The first part of this paper is an interim summary of the main results obtained with nitrogenous fertilizers in factorial nutritional studies at the G.C.R.I.; much of the data is as yet unpublished. The remainder of the paper is a review of recent information on the use of nitrogen for flower crops under glass, with particular reference to liquid feeding and to the levels of nitrate in the soil. This review is not complete in itself, but is concerned primarily with practical information from the U.S.A.; much of this information has been gathered from publications which are not always readily available in this country, or by correspondence with American workers. Far less information concerning the nutrition of flower crops is available from European countries, but reference is made to the practices of certain leading growers in southern England.

Some effects of nitrogen in nutritional studies at the G.C.R.I.

Carnations

Three factorial nutritional trials with carnations have been completed at the G.C.R.I., each lasting nearly two years. Solutions containing nitrogen and potassium were applied via Cameron irrigation equipment to beds of compost (soil + peat + grit) treated with various levels of superphosphate and lime. The results of the first trial have already been published[(34)]; the total number of blooms, averaged over all the other treatments, was highest at the intermediate of three nitrogen levels (125 ppm N), but in the limed plots there was some indication that the highest nitrogen level (225 ppm N) was the best.

Four levels of nitrogen (20, 90, 160, 230 ppm N) were compared in the second trial; the varieties grown were William Sim and Pink Sim, and 192 sub-plots were recorded. The total number of blooms increased significantly with nitrogen concentration up to 160 ppm N in the nutrient solutions, with a slight but insignificant decrease to 230 ppm N (Table 1, Trial II, A). When only those plots adequately supplied with phosphate are considered, the results (Table 1, Trial II, B) show the greatest number of blooms at 230 ppm N, though the increase from 160 to 230 ppm is not significant.

Table 1

Total blooms per sub-plot (18 plants) in trials II and III, averaged (A) over all other treatments and (B) at high P and K. Each value is the mean for two varieties (William Sim and Pink Sim)

Trial		p.p.m. N in nutrient solution			
		20	90	160	230
II	A	496	542	570	567
	B	509	581	630	641
III	A	403	519	511	494
	B	456	625	642	641

In the third trial of the series the greatest number of flowers, averaged over all other treatments, was found at 90 ppm N in the nutrient solutions, the decrease to 160 ppm N being slight but continuing to 230 ppm N (Table 1, Trial III, A). The difference between trials II and III in this respect is probably due to the increased severity of phosphorus and potassium deficiencies at the lowest levels of these nutrients during trial III, this in turn reducing the nitrogen requirement in some plots. When only those plots adequately supplied with P and K are considered (Table 1, Trial III, B), the highest numbers of blooms were obtained at 160 and 230 ppm N. These results do not indicate a clearly defined optimum concentration of nitrogen for maximum flower production, but in soil adequately supplied with phosphorus and potassium it appears that 160 ppm N is sufficient in this respect.

In the first of these nutritional studies with carnations[34] it was found that increasing levels of nitrogen significantly decreased the splitting of the calyx of the two varieties used (William Sim and Saugus White). This conclusion has been confirmed for William Sim and Pink Sim in trials II and III (Table 2).

Table 2

Percentage of blooms with split calyces

Trial	Variety	p.p.m. N in nutrient solution			
		20	90	160	230
II	William Sim	19·2	12·9	9·5	6·8
	Pink Sim	23·6	15·5	14·0	11·0
III	William Sim	12·7	8·0	6·2	4·7
	Pink Sim	17·5	10·5	9·4	7·2

Although the greatest response to nitrogen occurs at the lower concentrations, the beneficial effect of this nutrient continues at the higher concentrations.

Thus even the small decrease in percentage of 'splits' from 160 to 230 ppm N in trial II is statistically significant ($P=0 \cdot 05$); the overall effect of nitrogen on splitting is significant at $P=0 \cdot 001$. Although, at adequate levels of P and K, there was little difference in total bloom numbers between 160 and 230 ppm N (trials II and III), the small but significant decrease in splitting at the higher concentration suggests that concentrations above 160 ppm N are beneficial; 180–200 ppm N are therefore recommended in all waterings under favourable growing conditions.

Since, where trickle irrigation is used, nitrate tends to accumulate in the drier areas of the beds[34], some caution is necessary in suggesting optimal levels of soil nitrate. This difficulty was decreased in trials II and III, however, by doubling the number of irrigation rods to improve the distribution of water; the rods are now eight inches apart. The concentrations of nitrate-nitrogen in soil receiving the four nitrogen treatments, determined in the moist soil but reported on an oven-dry basis, are shown in Table 3; each value is the mean for four sampling dates. On this basis the recommended concentrations of 180 to 200 ppm N in the liquid feeds should correspond to about 125 to 150 ppm NO_3-N in the soil. Many factors may affect the nitrate content of the soil, however, including the distribution of nutrients by the irrigation system, the ease of drainage from the beds, the volume of solution applied at each watering, and possibly the source of nitrogen applied.

Table 3

The average nitrate concentrations (ppm NO_3-N on the basis of oven-dry soil) in trials II and III

Trial	p.p.m. N in nutrient solution			
	20	90	160	230
II	24	58	116	172
III	14	47	116	175

Winter lettuce

The 72 plots maintained for factorial studies of tomato nutrition (see *Rep. Glasshouse Crops Res. Inst.* from 1958 onwards) have also been used in alternation to study the nutrition of winter lettuce. The varieties grown were Cheshunt 5B in 1960–61 and Southdown in 1962–63. Apart from small amounts of nitrogenous fertilizer, the nutritional treatments depended entirely on residual fertilizers from the preceding tomato crops. The additional nitrogen consisted of 0, 1 and 2 oz of hoof and horn per sq. yd at the N1, N2 and N3 levels before planting in 1960, 1962 and 1963; in 1961 the only fertilizer treatment was 1 oz hoof and horn per sq. yd at the N2 level, since the N1 and N2 plots did not differ appreciably in nitrate content after removal of the tomato crop. The amounts of nitrogen applied to the preceding tomato crops are shown in Table 4.

Table 4

Total fertilizer–nitrogen applied per season in base and top dressings for tomatoes grown at three levels of nitrogen (oz N/sq. yd)

Year	Level of nitrogen N1	N2	N3
1958	0	0·75	2·25
1959	0	0·75	2·25
1960	0	1·00	2·50
1961 onwards	0·5	1·25	2·75

Effect of nitrogen on percentage of marketable lettuce

The percentages of marketable lettuce, averaged over all other combinations of nutritional treatment, are shown in Table 5.

Table 5

The main effect of nitrogen on the percentages of marketable lettuce 1960–63. Each value is the mean for 24 plots

Year planted	Level of nitrogen N1	N2	N3	P*
1960	97·6	97·2	78·8	0·001
1961	85·0	75·1	51·8	0·001
1962	72·4	58·5	36·6	0·001
1963	57·3	56·8	51·2	n.s.

* Significance of F-test after angular transformation.

Table 6

The effects of nitrogen, phosphorus and lime on the percentage of marketable lettuce from crops planted in 1962–63

Year planted	Lime	Phosphate	Level of nitrogen N1	N2	N3
1962	Ca 1	P1	28	7	1
		P2	93	91	58
	Ca 2	P1	56	30	15
		P2	96	96	93
1963	Ca 1	P1	20	5	0
		P2	93	82	50
	Ca 2	P1	54	47	50
		P2	58	93	99

The results in Table 5 show clearly that high levels of nitrogen can have an adverse effect on the lettuce crop, as in 1960–62. The data are somewhat misleading in this form, however, owing to interactions between nutrients. As an example of the complex nature of the results, the percentages of marketable lettuce from the crops planted in 1962 and 1963 are shown in Table 6, arranged factorially according to the levels of nitrogen, phosphorus and lime; each value is the mean of six plots (3K × 2Mg treatments).

The N3 level of nitrogen, and sometimes even the N2 level, reduced the proportion of marketable lettuce in the unlimed plots. A similar effect was found in the limed but phosphorus-deficient plots (Ca 2 P1) in 1962. In contrast, where both lime and phosphate were supplied, the highest (N3) level of nitrogen did not depress the percentage of marketable lettuce significantly in 1962, and over 99 per cent of the lettuce grown with high nitrogen under these conditions in 1963 were marketable (the difference between the N2 and N3 levels was not significant statistically). The N3 level of nitrogen was also satisfactory where lime and phosphate had been applied previously for tomatoes in 1960, and only in 1961 was there any indication that this level of nitrogen might be excessive under these conditions (88, 88 and 80 per cent marketable at N1, N2 and N3 respectively in the P2 Ca 2 plots).

Only in 1963 did the low (N1) level of nitrogen significantly reduce the percentage of marketable lettuce as compared with the N2 level in soil high in lime and phosphate (P=0·001; Table 6).

Effect of nitrogen on mean weight per lettuce

In the absence of added phosphate the N3 level of nitrogen always reduced the weight of the lettuce plants, and in three seasons out of four even the N2 level of nitrogen was apparently too high for lettuce under these conditions. At the high (P2) level of phosphate, however, the adverse effect of high nitrogen was lessened considerably, or even eliminated as in 1963, and the intermediate (N2) level of nitrogen was generally superior to the low (N1) level (Table 7).

Table 7

Effect of nitrogen on the mean weight (g) per lettuce harvested at the higher (P2) level of phosphate and the highest (K3) level of potash

Year planted	Level of nitrogen			L.S.D. (P=0·05)
	N1	N2	N3	
1960	70	73	55	± 8
1961	68	81	74	± 9
1962	70	80	68	±11
1963	54	91	90	±10

The N2 level of nitrogen increased the mean weight per lettuce as compared with N1 in 1961–63, the effect being significant in 1961 and 1963 and particularly marked in the latter year. In 1963 the highest (N3) level of nitrogen had no adverse effect in the presence of adequate phosphate and potash, but in the three preceding years the weight of the lettuce was adversely affected by high nitrogen; thus in 1960 and 1962 the weight per plant was significantly lower at

N3 than at N2. In 1960 the weight at N3 was even significantly lower than at N1; this result was unexpected, since at the N1 level the preceding tomato crop showed marked symptoms of nitrogen deficiency and no additional nitrogen was applied for the lettuce.

The data in Table 7 are based on both limed and unlimed plots, and the adverse effect of high nitrogen is partly due to acidity. In the unlimed plots supplied with phosphate the high (N3) level of nitrogen invariably had an adverse effect on the weight of the lettuce plants, as shown in Table 8.

Table 8

The effect of three levels of nitrogen on the weight (g) of lettuce plants grown with added phosphate in the absence of lime

Year planted	Level of nitrogen			L.S.D. (P=0·05)
	N1	N2	N3	
1960	59	65	47	± 7
1961	51	61	51	± 8
1962	61	62	45	± 9
1963	72	77	63	± 9

Even when soil acidity is eliminated by heavy liming, however, the response to nitrogen has generally been small and not always favourable. The mean weights of lettuce plants grown in the limed plots and adequately supplied with phosphate are shown in Table 9.

Table 9

The effect of three levels of nitrogen on the weight (g) of lettuce plants grown with added phosphate in limed soil

Year planted	Level of nitrogen			L.S.D. (P=0·05)
	N1	N2	N3	
1960	56	60	54	± 7
1961	62	68	58	± 8
1962	59	66	60	± 9
1963	44	93	100	± 9

In the first three years (1960–62) the intermediate level of nitrogen (N2) always gave the highest weight per plant in limed soil (Ca_2 P2), but the difference between N1 and N2 was not significant in any one of these years. The weight per plant at the high nitrogen (N3) level during 1960–62 never significantly exceeded that of plants grown at low nitrogen (N1), and was significantly lower than at the N2 level in 1961. The results for 1963 show a rather different pattern, however, the response to the N2 level of nitrogen being very marked; the N3 level also produced some further increase in weight of the plants grown in limed soil, though this was not significant statistically. As one possible explanation of the results of 1963 it may be noted that, under the more favourable combinations of treatments, the lettuce were larger than in the previous seasons.

Determinations of nitrate-nitrogen in the soil of all 72 plots were made during cropping (October–November) in 1961–63. The mean values at each level of nitrogen applied, averaged over the three seasons, were 11, 35 and 133 ppm NO_3-N (on the basis of oven-dry soil). It is tentatively suggested that levels of 30 to 60 ppm NO_3-N in the soil would generally be adequate for lettuce under our conditions, and that levels of 100 to 150 ppm NO_3-N would seem excessive. Nevertheless in 1963, when the lettuce grew better than in previous seasons of our trials, the heaviest plants, grown in the presence of high lime and phosphate, were obtained at the N3 level (127 ppm NO_3-N). It is thus difficult to generalize; the optimum nitrogen level doubtless depends on the extent to which other factors, both environmental and nutritional, limit crop growth.

Our results show little need for applying nitrogenous fertilizers for lettuce following tomatoes. Thus the N3 level of nitrogen applied to the preceding tomato crops is comparable with good commercial practice, and even in the absence of base nitrogen for the lettuce (as in 1961) the weight per plant was lower than at the N2 level; in the latter treatment 1 oz hoof and horn/sq. yd was applied for the lettuce, but the average nitrate level during cropping was only 39 ppm NO_3-N as compared with 156 ppm at N3.

The nitrogen uptake of the lettuce crop is relatively small. Calculations based on the 1962 crop show that the total nitrogen content of a 10-ton crop would be approximately 40 lb/acre (excluding roots), equivalent to the nitrogen supplied by 1 oz/sq. yd of hoof and horn.

Nitrogen and the tomato crop

The effects of nitrogenous fertilizers have been studied in factorial combination with phosphorus, potassium, magnesium and lime. Three levels of nitrogen were included, as shown in Table 4; from 1961 onwards the N1, N2 and N3 levels were equivalent to 151, 378 and 832 lb N/acre per season, supplied partly as hoof and horn in the base and partly as nitro-chalk in the top dressings. Of these levels, N1 was intended to show nitrogen deficiency, N2 was sub-optimal though free from marked deficiency symptoms, and N3 was considered to supply a fully adequate level of nitrogen for the most favourable combinations of the other nutrients. At low levels of phosphate the N3 level of nitrogen proved somewhat excessive as judged by the extremely dark green appearance of the plants. Where adequate levels of the other nutrients were maintained, the inclusion of still higher levels of nitrogen would be necessary before the optimum could be determined with certainty. There is, however, little or no indication that plants grown at the N3 level of nitrogen would have benefited from still higher levels of this nutrient even under the most favourable conditions. The quantity of nitrogen supplied at the N3 level (832 lb N/acre from 1961 onwards) is very similar to that given in Advisory Leaflet 520[21], and further discussion of rates of application seems unnecessary here.

The levels of soil nitrate in this trial with solid fertilizers, averaged over the last two seasons for samples taken between planting and June to July, were 12, 35 and 156 ppm NO_3-N. In an earlier experiment in which nitrogen and potash were applied in solution via a hose (see the following section on liquid feeding of tomatoes), the most favourable combination of yield and quality was found in plots which averaged about 120 ppm NO_3-N. Concentrations of 100 to 150 ppm NO_3-N during cropping thus appear to be suitable for tomatoes grown under our conditions.

Liquid feeding of tomatoes

No factorial studies on nitrogen concentrations in liquid feeds for tomatoes have yet been made at the G.C.R.I. In other liquid feeding trials with this crop, where nitrogen was not an experimental factor but was varied according to the appearance of the plants, experience over many seasons has shown that concentrations of 175 to 225 ppm N in the diluted nutrient solutions appear to give good growth and colour under most conditions.

In an experiment started in 1957, four concentrations of a liquid feed supplying nitrogen and potassium at a K_2O/N ratio of 1·75 were compared for three years(35). The best combination of yield and fruit quality at this K_2O/N ratio occurred at or slightly below 225 ppm N with 394 ppm K_2O. The somewhat dark green appearance of the plants suggested that a slightly lower nitrogen concentration, and a correspondingly higher K/N ratio, would have been preferable.

Advisory Leaflet 520 calls for the use of a 'high nitrogen' feed (315 ppm N) for main crop tomatoes in June and July, when fruit is setting on the fourth truss; for varieties of the Potentate type, or for other varieties if planted early, this treatment is started three weeks earlier. Although this point has not been investigated at the G.C.R.I., the appearance of our plants generally suggests the need for additional nitrogen at about this stage. Practice at the G.C.R.I. differs somewhat from Advisory Leaflet 520 in that high nitrogen feeds are often introduced even earlier (early May or late April), but at a lower concentration (250 to 275 ppm N) and usually for not more than three weeks.

Various sources of nitrogen have been compared in liquid feeds for tomatoes at the G.C.R.I.; the concentrations of nitrogen and K/N ratios were the same for all treatments at any one time, but were varied according to the stage of growth. For the first two seasons of the trial the four sources of nitrogen were urea (organic-N), ammonium sulphate (NH_4-N), ammonium nitrate (NH_4-N+NO_3-N) and a mixture of potassium and sodium nitrates (NO_3-N); potassium for the first three of these treatments was supplied as potassium sulphate. In 1960, ammonium nitrate gave the highest yield, though not significantly greater than ammonium sulphate or urea, and potassium + sodium nitrates gave the lowest yield ($P=0·01$). In 1961, ammonium sulphate gave the highest yield (significantly greater than either ammonium nitrate or urea, $P=0·01$), and potassium + sodium nitrates again gave the lowest yield. In both these seasons ammonium sulphate gave the lowest percentage of unevenly ripened fruit, closely followed by ammonium nitrate, whereas urea and the 'all-nitrate' feed gave a higher proportion of unevenly coloured fruit. Salinity measurements made on samples taken midway between the trickle nozzles showed that fruit quality improved with increasing salinity. The differences in fruit quality (though not in yield) were thus not necessarily due to the chemical form of the nitrogen supplied (i.e. NH_4-N or NO_3-N), but could be due partly to physical effects. It was also uncertain whether the adverse effect of the 'all-nitrate' feed was due to the form of the nitrogen or to the inclusion of sodium ions.

In 1962–63, while retaining the ammonium sulphate and the potassium + sodium nitrate treatments, further attention was given to the nature of the cations present in the 'all-nitrate' feeds. Sodium salts were found to reduce the yield, but when calcium was substituted for sodium (i.e. potassium + calcium nitrates) the proportion of irregularly ripened fruit was significantly higher than for ammonium sulphate + potassium sulphate.

To summarize these results with various nitrogen sources for tomatoes, the best yields and quality (by visual grading) were obtained with ammonium sulphate and with ammonium nitrate, despite the extreme acidity induced in the soil; the pH values immediately below the nozzles were 4·3 and 4·5 respectively, though the affected areas were localized. In view of the danger from blossom-end rot under acid (calcium-deficient) conditions it would perhaps seem inadvisable to recommend the 'all ammonium-N' feeds commercially at present except in highly calcareous soils.

A review of some recommendations on the use of nitrogen for flower crops

The object of this section is to bring together various recommendations for the liquid feeding of commercial flower crops and for the optimal levels of nitrate in the soil. The data have been obtained mainly from American sources, though the practices and recommendations of some English growers and horticultural experts have also been included. The emphasis in this review, which is not claimed to be complete, is primarily on the practical and commercial aspects of the subject. In the nutrition of flower crops it indeed seems that the grower has often been ahead of the adviser, and that the latter has in the past received relatively little guidance from the research worker. Studies of the nutrition of flower crops in progress at the G.C.R.I. should help to remedy this situation, but no references to this work have been included in the present review.

Published recommendations for soil nitrate (or ammonium-nitrogen) levels for flower crops are almost entirely American in origin, and commonly based on the Spurway system of soil testing. Where the Spurway system is not used, the results of American experience have generally been ignored. The basis of the Spurway tests for soil nitrate is therefore discussed in an attempt to utilize the available recommendations.

Liquid feeding of glasshouse flower crops

Carnations. Holley and Woodward[16] compared the growth of carnations at four concentrations of a liquid feed supplying N, K and other elements. The solution regarded as 'standard' contained 131 ppm N and 100 ppm 'potash', and was compared with stronger ($\times$ 1·25 = 163 ppm N) and weaker ($\times$ 0·75 = 98 ppm N and $\times$ 0·5 = 65 ppm N) solution sof the same ingredients. The strongest solution (163 ppm N) gave the highest 'mean grade', but not significantly higher than the 'standard' solution. The 'standard' solution (131 ppm N) gave the highest number of blooms, but not significantly higher than at 163 ppm. The two weaker solutions gave less satisfactory results.

In a panel discussion reported in the *Maryland Florist*[1], 6 to 8 oz of 25–5–20 fertilizer per 100 (U.S.) gallons were recommended for carnations, corresponding to 113 to 150 ppm N.

Davidson[7] proposed a solution containing 3·75 lb 20–20–20 soluble fertilizer, 1·25 lb potassium chloride, 1·25 lb sodium nitrate and 0·5 lb Epsom Salts per 1,000 (U.S.) gallons of water; this corresponds to approximately 114 ppm N.

Many of the current recommendations for the liquid feeding of carnations in the U.S.A. are based on the work of Eck at Massachusetts. Eck, Campbell and Spelman[9] concluded from a factorial study that 180 ppm N and 200 ppm K gave the best growth and flower quality; the results reported for number of blooms per plant and for mean quality index showed no statistically significant differences between the nitrogen treatments, but the dry matter content per plant increased with nitrogen level from 40 to 360 ppm in the liquid feeds. Other workers refer to the work of Eck at Massachusetts, but give formulae supplying 200 ppm N and K. Four such formulae were given by Gianfagna and Nichols[11]:

A	6 oz NH_4NO_3 + 8 oz KNO_3	each per 100 gallons (U.S.)
B	10·5 oz $NaNO_3$ + 8 oz KNO_3	
C	10·5 oz $Ca(NO_3)_2$ + 8 oz KNO_3	
D	4 oz urea + 8 oz KNO_3	

For use with calibrated dilution equipment, these quantities per gallon, multiplied by 2·4 (which includes conversion from U.S. to Imperial gallons), could be applied at a dilution of 1 in 200. The authors suggest that the alternative formulae should be used at intervals to help maintain pH and soil structure.

Holley and Baker[15] note that 'Most commercial growers have found that the C.S.U. (Colorado State University) liquid feeding rate is not adequate to keep soil tests as high as they would like. This rate has concentrations of N and K at approximately 130 ppm'. This comment must refer to earlier practice at Colorado, since the amounts of fertilizer-grade chemicals per 1,000 gallons of water 'at present' are given as 4 lb ammonium nitrate, 0·25 lb sodium nitrate, 2 lb potassium chloride, 0·5 lb Epsom Salts and 0·5 oz (winter) or 1 oz (summer) of borax; this corresponds to 163 ppm N assuming 33 per cent N in the commercial ammonium nitrate (cf. the lower concentration of 131 ppm N quoted by Holley and Woodward[16]).

The current recommendation from Fernhurst[20] is for the use of equal parts of 'High Potash' and 'Standard' 'Solufeed' at 3 lb per 250 gallons, giving 236 ppm N (283 ppm K_2O); the frequency of application would vary somewhat according to analysis, but would generally consist of two applications of nutrient solution to one application of plain water.

Butters[6] recommends the use of a liquid feed containing 200 ppm N and 200 ppm K_2O at every watering, except when modified in the light of soil analysis at 6–8 week intervals.

Chrysanthemums. For 'All the year round' chrysanthemums, Searle and Machin[26] recommended feeding as follows at all waterings:

Low light (winter) and whenever buds showing	45–90 ppm N
General	60–120 ppm N
Bright summer weather	90–180 ppm N

In a summary of a panel discussion in the *Maryland Florist*[1], 6 to 8 oz of 25–10–10 fertilizer per 100 gallons (U.S.) were recommended for chrysanthemums at every watering; this corresponds to 113 to 150 ppm N.

The feeds proposed by Framptons Nurseries Ltd.[10], based on potassium nitrate and urea and applied at alternate waterings, correspond to 243 ppm N under high light intensity or for early growth generally, and to 186 ppm N under lights to flower in the first four months of the year.

For 'All the year round' chrysanthemums, Butters[6] considers that no feeding is generally necessary provided that a base fertilizer, decided after analysis but commonly including about 4 oz hoof and horn per sq. yd, is applied after steaming and flooding. If nitrate levels are very low at planting time, however, one or two high nitrogen feeds should be given in the first week or two after planting. For chrysanthemums in pots (J.I.2), liquid feeding at alternate waterings is recommended, starting two weeks after potting and continuing until the buds show colour. From April to September the diluted nutrient solution (potassium nitrate and urea) contains 360 ppm N and 220 ppm K_2O; from October to March the nitrogen concentration is reduced to 280 ppm N.

Roses. In a panel discussion in the *Maryland Florist*[1], 4 to 6 oz of 25–10–10 fertilizer per 100 gallons (U.S.) were recommended, corresponding to 75 to 113 ppm N.

Davidson[8] recommended 2·8 lb of 20–20–20 soluble fertilizer and 1·2 lb calcium nitrate per 1,000 gallons (U.S.) of water, applied at 3 pints per sq. yd of bench or bed whenever watering was required; calculation shows a nitrogen concentration of 90 ppm N.

General recommendations for flower crops

For pot plants Shanks[29] recommended the following amounts of 25–10–10 fertilizer per 100 gallons:

3 lb	every 2 weeks	(= 900 ppm N)
1½ lb	every week	(= 450 ppm N)
3 oz	every day	(= 56 ppm N)

Recommendations for various flower crops by *Maryland Florist*[1], expressed in ounces of 25–10–10 fertilizer per 100 gallons, are:

Annuals	3	(57 ppm N)
Azalea	2+	(38 ppm N)
Cyclamen	3	(57 ppm N)
Gardenia	4–6	(75–113 ppm N)
Geranium	4–6	(75–113 ppm N)
Hydrangea—summer	2	(38 ppm N)
Hydrangea—forcing	6–8	(113–150 ppm N)
Poinsettia	6–8	(113–150 ppm N).

Holley[13] states that in most cases a standard nutrient solution used in the U.S.A. will contain about 125–170 ppm of N, a similar amount of K_2O, and somewhat less P_2O_5.

Information supplied by Kiplinger[18] refers to experiments at the Ohio State University in which roses, carnations, chrysanthemums, cyclamen, hydrangeas, lilies and poinsettias were grown in soil (some also in sand and peat) using 20–20–20 and 25–5–20 fertilizer at 1 oz per 10, 15, 20, 30 and 60 gallons of water. The plants grown with 1 oz per 10 gallons of nutrient solution were best, closely followed by those receiving 1 oz per 15 gallons. Plants grown at 1 oz per 20 gallons were somewhat less desirable, and those grown at 1 oz per 30 or 60 gallons were starved. One ounce of fertilizer containing 20 and 25 per cent N per 10 gallons (U.S.) corresponds to 148 and 185 ppm N respectively. Reference is also made to the work on carnations at Massachusetts (presumably Eck and co-workers), and various formulae for preparing liquid feeds supplying 200 ppm each of nitrogen and 'potassium' after dilutions are tabulated; from the formulae given it appears that the diluted solutions would contain 200 ppm K_2O rather than K. It is stated that these nutrient solutions should be satisfactory for a wide variety of flower crops if the soil or other medium in which they are grown is watered adequately; should soil tests reveal a high salt content, the watering should be more thorough so as to cause leaching, or the concentration of the nutrients may be reduced to 150 ppm.

The Spurway system of soil testing

This system of soil testing is widely used in the U.S.A., and the levels of soil nutrients recommended for various crops, particularly the flower crops, may be found in numerous publications ranging from scientific papers to advisory notes. Surprisingly little use of this information appears to have been made in this country, probably owing to uncertainties concerning the procedures used and the mode of expression of the results. It is in fact recognized by soil scientists in America that the situation with regard to this system is somewhat confusing, but that nevertheless it is capable of giving useful information and has become accepted by the growers.

In order to use the considerable information available from Spurway test results it is necessary to establish various details of the procedure, particularly the ratio of extractant to soil. Since nitrates are highly water-soluble and relatively little adsorbed, the nature of the extractant (dilute acetic acid) need not concern us here. Spurway and Lawton(31) specified 'one level spoonful ($\frac{1}{2}$ teaspoon)' of soil in a glass tube, with water added to the 13 ml mark; one drop of acetic acid was added and the tube shaken for one minute. The results are expressed as ppm NO_3 in the extract. The soil was apparently to be used without preliminary air-drying.

Numerous modifications of this extraction procedure have been developed in America. Kiplinger(18) states that the U.S. teaspoon, as used in the Ohio State University laboratories, has a capacity of 4·9 ml; 26 ml of extractant are added to one level teaspoon of air-dry soil. Davidson(8) states that a half-teaspoon hold abouts 2·1 g of air-dried greenhouse soil, giving an extractant/soil ratio of about 6:1, but that some laboratories run the Spurway tests at a 5:1 ratio. At Cornell(5) the test is made with 5 g of air-dry soil and 25 ml of extractant (Spurway), except for highly organic soils when one level teaspoon is taken. To overcome differences in the size of teaspoons, the half-teaspoons used at Michigan State University, where the Spurway system originated, are now calibrated to deliver 2·5 g of soil(30).

Owing to variations in the bulk density of horticultural soils, the choice of soil volume or soil weight at different testing centres, and to a lesser extent to other uncertainties such as moisture content of soil and size of teaspoon, it is not possible to express the result of this test precisely as ppm in the soil. Approximate factors can, however, be deduced so as to allow comparison with other testing procedures. Thus from Spurway and Lawton(31), 1 ppm in the extract is equivalent to 8 lb/acre. If the weight of one acre of soil to a depth of six inches be taken as approximately 2×10^6lb, then 1 ppm NO_3 in the extract corresponds to 0·9 ppm NO_3-N in the soil. Using 5 g of soil with 25 ml of extractant[5], 1 ppm NO_3 in the extract is equivalent to 1·13 ppm NO_3-N in the soil. Calibration of a half-teaspoon to hold 2·5 g of soil[30] would give results in which 1 ppm NO_3 in the extract is equivalent to 1·17 ppm NO_3-N in the soil. Still higher factors would result where soils of low bulk density were tested on a volume basis. For practical purposes, however, it would seem possible to regard figures for nitrate quoted on the Spurway system as approximately equivalent to ppm NO_3-N in the soil; any error in this assumption is likely to be in the direction of under-estimation.

Published recommendations on soil nitrate levels for flower crops

Most of the recommendations in this section are based on the Spurway system. All values obtained by this procedure are marked 'Spurway' in the text and tables; these values refer to ppm NO_3 in the soil extract but, as discussed in the preceding section, it is suggested that they may be interpreted as approximating to ppm NO_3-N in the soil. This is undoubtedly an over-simplification, but the errors involved in this assumption are unlikely to exceed the uncertainties inevitably associated with the designation of an optimum level of soil nitrate for the various crops. The results quoted from Weinard[32] are expressed in lb NO_3-N per acre; for approximate comparison with other workers the corresponding ppm NO_3-N in the soil has been added in brackets, assuming a weight of 2×10^6lb per acre to a depth of six inches.

Carnations. In a survey of commercial carnation houses Weinard[32] found generally 50–150 lb NO_3-N per acre (25–75 ppm NO_3-N in the soil), and recommended 100–150 lb N per acre (50–75 ppm N in the soil). Beach and Mussenbrook[3] attempted to maintain five levels of nitrate from 0–100 ppm (Spurway) in the soil. The levels achieved were considerably lower than intended, and the optimum level of nitrate appeared to be higher than any of the treatments, the highest of which averaged 33–45 ppm (Spurway). Post[24] proposed 25–150 ppm (Spurway), suggesting a wide tolerance range. Boodley[4], in a manual prepared with the co-operation of the New York State Flower Growers Association, recommended 25–100 ppm (Spurway). Holley and Baker[15] advised 40–100 ppm (Spurway). The recommendations from Rutgers State University[8] designate the range 25–100 ppm NO_3-N in the soil as 'good' (see Table 10), together with 3–30 ppm NH_4-N. Values greater than 200 ppm NO_3-N and 60 ppm NH_4-N are rated as 'excessive'. Butters[6] recommends maintaining 60–80 ppm NO_3-N in soil extracts prepared at a 5:1 ratio (Morgan). This corresponds to a higher level of soil nitrate (300–400 ppm NO_3-N) than in any of the preceding references. Recommendations for carnations from Cornell[22] suggest 25–60 ppm (Spurway).

Chrysanthemums. Post[24] recommended a soil nitrate level of 10–50 ppm (Spurway) for chrysanthemums, a range of values considerably lower than that for carnations (Table 10). The detailed recommendations from Rutgers State University[8] are the same as those given for carnations; levels of 25–100 ppm NO_3-N and 3–30 ppm NH_4-N are designated 'good' (see also other categories in Table 10). The Cornell recommendation[22] is 25–60 ppm (Spurway), as for carnations.

Table 10

Levels of soil nitrate and ammonium recommended for carnations

Author	Recommendation		
Weinard [32]	Found commercially 25–75 p.p.m. NO_3–N* Recommended 50–75 ,, ,,		
Beach and Mussenbrook [3]	Optimum > 33–45 p.p.m. (Spurway)		
Holley *et al.* [14] Holley and Baker [15]	40–100 p.p.m. (Spurway) 40–100 ,, ,,		
Post [24]	25–100 p.p.m. (Spurway)		
Boodley [4]	25–100 p.p.m. (Spurway)		
Davidson [8]	p.p.m.	NO_3–N**	NH_4–N**
	Poor	< 15	< 1
	Fair	15–25	1–3
	Good	25–100	3–30
	High	100–200	30–60
	Excessive	> 200	> 60
Butters [6]	300–400 p.p.m. NO_3–N***		
New York State College of Agriculture [22]	25–60 p.p.m. (Spurway)		

*Originally given in pounds of nitrate-nitrogen per acre and here converted to p.p.m. NO_3–N assuming a weight of 2 × 10^6 lb. per acre to a depth of six inches.

**Expressed as p.p.m. N in the air-dry soil; determined in 0·2 N-sodium acetate extract (pH 5·0) at an extractant/soil ratio of 2:1.

***Originally given as 60–80 p.p.m. NO_3–N in soil extract at 5:1 (Morgan).

The Spurway values are p.p.m. NO_3 in the extract, but may be taken as approximately equivalent to p.p.m. NO_3–N in the soil.

Roses. In a study of the nutrient status of commercial rose beds Weinard (1938) found 50–200 lb NO_3—N per acre (approximately 25–100 ppm NO_3—N in the soil). Weinard recommended 150–200 lb N per acre (75–100 ppm NO_3—N in the soil). Seeley[27] concluded that the minimum desirable level of nitrate for roses was between 10 and 25 ppm (Spurway), possibly 15–18 ppm (Spurway). Flower production increased up to the highest nitrogen level in this trial (50 ppm Spurway), and no upper limit could therefore be deduced. Post and Howland (1946) found that total blooms per plant increased with the level of nitrate over the range 10–100 ppm (Spurway). Maximum economical production was thought to occur between 25 and 75 ppm (Spurway), as 100 ppm (Spurway) was considered too expensive to maintain since it required weekly application of fertilizer. Soil analyses at weekly intervals were also thought to be necessary to avoid damage from over-fertilization where high nitrate levels were maintained. Seeley and Post[28] advised maintaining levels of 25–100 ppm (Spurway), their recommendations being based on three sets of experiments[25,27,28]. The production of saleable blooms was reduced at 150 ppm

(Spurway), and the plants were damaged at 300 ppm (Spurway). In the recommendations from Rutgers State University[8], levels of 50–150 ppm NO_3-N and 3–30 ppm NH_4-N are designated as 'good'; for further details see Table 11. The recommendation from Cornell[22] is for 25–100 ppm (Spurway).

Table 11

Levels of soil nitrate and ammonium recommended for roses.
(*See footnotes to Table* 10)

Author	Recommendation		
Weinard [32]	Found commercially	25–100 p.p.m. NO_3-N*	
	Recommended	75–100 ,, ,,	
Seeley [27]	Minimum 10–25 p.p.m. (Spurway)		
Post and Howland [25]	Maximum economic level 25–75 p.p.m. (Spurway)		
Seeley and Post [28]	Recommendation	25–100 p.p.m. (Spurway)	
	Reduction of saleable blooms	150 p.p.m. ,,	
	Plant injury	300 p.p.m. ,,	
Davidson [8]	p.p.m.	NO_3-N**	NH_4-N**
	Poor	<25	<1
	Fair	25–50	1–3
	Good	50–150	3–30
	High	150–250	30–60
	Excessive	>250	>60
New York State College of Agriculture [22]	25–100 p.p.m. (Spurway)		

Other Flower Crops. Numerous recommendations on the optimum levels of soil nitrate for flower crops have been given by American workers, mainly based on the Spurway system. A selection of the data available is given in Table 12. Whilst it seems unlikely that all these crops have been fully examined in nutritional trials, there is little doubt that the recommendations embody a great deal of practical experience.

It is not proposed to discuss these recommendations in detail. The minimum soil nitrate levels indicated by Post[24] tend to be a little lower than the others listed. The recommendation that soil nitrate levels for Sweet Peas should not exceed 10 ppm (Spurway) is associated with symbiotic nitrogen fixation. From the Cornell data[22] it would seem that 20–40 ppm (Spurway) would suit most of the crops mentioned. The general recommendations from Ohio are for rather higher levels of soil nitrate. It is specifically stated, however, that the Ohio data do not apply where nitrogen is applied in solution at all waterings (see final section of this paper). These general recommendations from Ohio are also probably intended to cover not only crops such as referred to by other authors in Table 12 but also the major crops referred to in the preceding sections. The distinction made in the Ohio recommendations between plants with restricted and unrestricted root-run is of particular interest.

Table 12

Levels of soil nitrate recommended for various flower crops.
(See footnotes to Table 10)

Author	Recommendation			
Weinard [32]	Gardenia	Found 13–75 p.p.m. NO_3–N* Recommended 50–75 p.p.m.		
Jackson [17]	Geranium	20–50 p.p.m. (Spurway)		
Post [24]	Antirrhinum	10–50 p.p.m. (Spurway)		
	Camellia	15 „ „		
	Euphorbia	> 10 „ „		
	Sweet Pea	< 10 „ „		
New York State College of Agriculture [22]	Antirrhinum	15–30 p.p.m. (Spurway)		
	Cyclamen	25–40 „ „		
	Dahlia	20–30 „ „		
	Gardenia	25–40 „ „		
	Geranium	25–40 „ „		
	Gladiolus	10–15 „ „		
	Hydrangea (field)	20–40 „ „		
	„ (forced)	30–50 pink or white)		
	„ „	20–30 (blue)		
	Lilies	30–50 p.p.m. (Spurway)		
	Poinsettias	25–40 „ „		
	Saintpaulias	20–30 „ „		
	Stocks	25–40 „ „		
Davidson [8]	Azaleas	p.p.m.	NO_3–N**	NH_4–N**
		Poor	< 10	< 4
		Fair	10–20	4–10
		Good	20–75	10–50
		High	75–125	50–75
		Excessive	> 125	> 75
Davidson [8]	Bedding plants	p.p.m.	NO_3–4	NH_4–N**
		Poor	< 10	< 1
		Fair	10–25	1–3
		Good	25–75	3–30
		High	75–150	30–60
		Excessive	> 150	> 60
Ohio Florists' Association [2]	Young plants and cuttings.	20–50 p.p.m. (Spurway)		
	Established plants in beds with bottoms.	75–90 p.p.m. (Spurway)		
	Ground beds (roots unrestricted).	40–60 p.p.m. (Spurway)		

Note: The Ohio recommendations do *not* apply where continuous liquid feeding is practised.

Sources of nitrogen for flower crops

The nitrogenous fertilizers listed in the various American recommendations for liquid feeding include ammonium sulphate, ammonium nitrate, potassium nitrate, sodium nitrate, calcium nitrate and urea. The four formulae given by

Gianfagna and Nichols[11] include two supplying nitrogen entirely as nitrate. The eight formulae listed by the Ohio Florists' Association also include two 'all-nitrate' feeds, together with one in which all the nitrogen is supplied as ammonium sulphate and another in which urea is the sole nitrogen-source. Holley[12], however, found differences in the growth of young carnations in sand culture when supplied with six alternative sources of nitrogen. Plants receiving ammonium, calcium or sodium nitrate produced the greatest fresh weight. The percentage dry matter content of plants receiving the sodium salt was low, however, and this was thought likely to result in poor quality flowers. Ammonium sulphate was the least satisfactory source of nitrogen as judged by both the fresh and dry weights of the plants.

White and Goddard[33] found that urea-formaldehyde, a slow-release fertilizer, compared favourably with ammonium nitrate as a source of nitrogen for carnations. Our own laboratory investigations of the properties of urea-formaldehyde compounds were not entirely encouraging, however[19]. Fertilizers intended to release nutrients slowly, whether biologically or physically (e.g., by surface coating[23]); suffer from the possible disadvantage that little or no control of the nutrient status of the plants is possible after application; the rate of release of nutrients may be affected by numerous factors, and will not necessarily be suited to the requirements of the crop at all times. In a leaflet prepared for the Ohio Florists' Association[2] it is stated that '. . . suggestions on fertilization are difficult to make when they (slow-release or coated fertilizers) have been used. The materials themselves cannot be leached from the soil and overdoses may seriously injure the roots and cause irreparable damage to the crop'. White and Goddard[33] found no damage to plants, however, when 4 lb urea-formaldehyde/100 sq. ft (5·8 oz/sq. yd) were applied either before planting or to the surface around growing plants. The use of 'slow-release' fertilizers (magnesium ammonium phosphate, coated fertilizers, various commercially available urea-formaldehyde compounds and urea-crotonaldehyde) in peat/sand mixtures is being investigated by Bunt at the G.C.R.I. (see Annual Report for 1963, pp. 105–6). Promising results have been obtained with a commercial magnesium ammonium phosphate preparation (analysis 7–40–4) at 8 oz/bushel, and with certain of the coated fertilizers. Incubation tests suggest, however, that the release rates are not as low as has sometimes been thought.

The problem of assessing the nitrogen requirements of glasshouse crops

It is becoming increasingly accepted that continuous liquid feeding may not maintain soil nutrient levels as high as would result from comparable amounts of fertilizer applied periodically in solid form. This point has already been noted in the recommendations of the Ohio Florists' Association (Table 12). Holley and Woodward[16] noted that, where liquid feeding was practised, 'regular soil tests also revealed that nitrates and potash were difficult to maintain at levels customarily sought when fertilizers are applied dry; . . . with well drained soil and constant liquid feeding, is there not a better way to determine the nutritional status of a plant than soil testing?'

In my experience the value of soil analysis prior to planting, and also during cropping where top dressings are applied either in solid form or uniformly by hose watering, is unquestionable. The distribution of nutrients from some types of low-level irrigation system needs further consideration, however. With trickle irrigation equipment of the type widely used for tomatoes it is even debatable whether soil analysis during cropping has any general application whatsoever. It seems likely that soil analyses made during cropping, particularly for highly soluble nutrients such as nitrate, may require somewhat differing scales of interpretation according to the mode of fertilizer application. Plant analysis may assist with this problem, but suffers from the obvious disadvantage that it refers to nutrient status at some previous stage of growth rather than to the current position.

It is suggested that for trickle irrigation the proper control of nutrient status during cropping should be based on detailed knowledge of the amounts of fertilizer applied, checked periodically by soil or plant analyses as necessary. For this to be fully successful, however, would require more precise recording of the volume and concentration of liquid feeds applied than is often found at present in commercial practice.

References

1. ANON. (1960). Liquid fertilization. *Maryland Florist*, No. 67, 3–4.
2. ANON. (1964). *Interpreting soil tests*. Ohio Florists' Association, Columbus, Ohio.
3. BEACH, G. and MUSSENBROOK, A. (1948). Effect of nitrogen and potash on Patrician carnations in soil. *Proc. Amer. Soc. Hort. Sci.*, **52,** 487–9.
4. BOODLEY, J. W. (1961). *Carnations; a manual of the culture, insects and diseases and economics of carnations*. Cornell University, Ithaca, New York.
5. BOODLEY, J. W. (1964). Private communication.
6. BUTTERS, R. E. (1964). Private communication (A. G. Sparkes Ltd., Angmering, Sussex).
7. Davidson, O. W. (1961). *Fertilizer/water for carnations*. Rutgers University (mimeograph).
8. DAVIDSON, O. W. (1964). Private communication.
9. ECK, P., CAMPBELL, F. J. and Spelman, A. F. (1962). Effect of nitrogen and potassium fertilizer on soil concentrations and on production, quality and mineral composition of carnations. *Proc. Amer. Soc. Hort. Sci.*, **80,** 565–70.
10. Frampton's Nurseries Ltd., (1962). *Chrysanthemum flowering from direct planted cuttings*. 7th Edit. Chichester, Sussex.
11. GIANFAGNA, A. and NICHOLS, L. P. (1960). Commercial carnation production. *Carnation Craft*, No. 53, 2–5.
12. Holley, W. D. (1956). *Further investigations on the effects of nitrogen sources on carnation growth*. Colorado Flower Growers' Association, Bull. 85.
13. HOLLEY, W. D. (1964). Private communication.
14. HOLLEY, W. D., WAGNER, D. L. and FARMER, R. (1951). The response of carnation varieties William Sim and White Patrician to various levels of nitrate and soil moisture. *Colorado Flower Grower's Association, Bulls.* **26, 27**; ref. from Holley and Baker (1963).
15. HOLLEY, W. D. and BAKER, R. (1963). *Carnation production*. W. C. Brown Co., Dubuque, Iowa.
16. HOLLEY, W. D. and WOODWARD, C. W. (1958). Response of carnations to different rates of liquid feeding. Colorado Flower Growers' Association, Bull. 104.
17. JACKSON, H. (1947). How geraniums respond to different nutrient levels. *Florists' Exchange*, **109,** ref. from Post (1956).
18. KIPLINGER, D. C. (1964). Private communication.
19. LONG, M. I. E. and WINSOR, G. W. (1960). Isolation of some urea-formaldehyde compounds and their decomposition in soil. *J. Sci. Food Agric.*, **11,** 441–5.
20. LOWCOCK, A. M. (1964). Private communication (Plant Protection Ltd., Fernhurst, Surrey).
21. Ministry of Agriculture, Fisheries and Food. *Liquid Feeding of Tomatoes*. Advisory Leaflet 520.
22. New York State College of Agriculture (1964). *Cornell recommendations for commercial floriculture crops*. Ithaca, New York.
23. OERTLI, J. J. and LUNT, O. R. (1962). Controlled release of fertilizer minerals by incapsulating membranes. Parts I and II. *Proc. Amer. Soc. Soil Sci.*, **26,** 579–87.

24. POST, K. (1956). *Florist Crop Production and Marketing*. Orange Judd Publishing Co., New York.
25. POST, K. and HOWLAND, J. E. (1946). The influence of nitrate level and light intensity on the growth production of greenhouse roses. *Proc. Amer. Soc. hort. Sci.*, **47**, 446–50.
26. SEARLE, S. A. and MACHIN, B. J. (1957). *Chrysanthemums the year round*. Blandford Press, London.
27. SEELEY, J. G. (1943). The effect of various nitrate levels on the growth and production of greenhouse roses. *Proc. Amer. Soc. Hort. Sci.*, **42**, 629–34.
28. SEELEY, J. G. and POST, K. (1948). Soil nitrate levels for roses. *Proc. Amer. Soc. Hort. Sci.*, **51**, 613–7.
29. SHANKS, J. B. (1958). Using fertilizers effectively. *Maryland Florist*, No. 50.
30. SHICKLUNA, J. C. (1964). Private communication.
31. SPURWAY, C. H. and LAWTON, K. (1949). Soil Testing. *Michigan Agricultural Experiment Station. Tech. Bull.* 132.
32. WEINARD, F. F. (1938). Some results with rapid tests on rose, carnation and gardenia soils. *Proc. Amer. Soc. Hort. Sci.*, **35**, 855–7.
33. WHITE, H. E. and GODDARD, B. (1957). The effect of nitrogen from two sources on carnation growth. *Massachusetts Flower Growers' Association*, Bull. 43.
34. WINSOR, G. W. and LONG, M. I. E. (1962). A factorial nutritional study of the perpetual-flowering carnation grown under glass. *J. Hort. Sci.*, **37**, 299–312.
35. WINSOR, G. W., DAVIES, J. N., MESSING, J. H. L. and LONG, M. I. E. (1962). Liquid feeding of glasshouse tomatoes; the effects of nutrient concentration on fruit quality and yield. *J. Hort. Sci.*, **37**, 44–57.

Discussion

Mr. R. R. Charlesworth noted that *Dr. Winsor* recommended a concentration of nitrogen in the liquid feed for carnations of 180 to 200 ppm. He wondered whether levels as high as these were justified because many growers in the south-east had excessive levels of nitrate in their soils and restriction of nitrogen feeding had to be recommended. *Mr. Charlesworth* asked what the soil pH values were in the lettuce experiment with the different nitrogen regimes. He also asked, in view of what *Mr. Kirkby* had said about the effects of ammonium nutrition on the organic acid content of plants, whether this would be an important factor in the growing of the tomato crop. In answer to the first question, *Dr. Winsor* said that if the nitrate content of the soil became high he would use this as a guide rather than continue with the high level of nitrogen feeding. He said that on the limed plots in the lettuce experiment the soil pH values were all above 7·0. On the un-limed plots the pH values ranged from about 7·0 on the low nitrogen plots, to between 4·5 and 5·0 on the high nitrogen plots. Referring to the question on tomato nutrition *Dr. Winsor* said that in experiments carried out at the Glasshouse Crops Research Institute there were no striking differences in the acidities of fruit from the various treatments testing different nitrogen sources. He further said that the salinity of the soil would also affect the acidity of the fruit and this would be a complicating factor. *Mr. W. Dermott*, asked *Dr. Winsor* whether growers in the U.S.A. were carrying out their own soil nitrate analyses. *Dr. Winsor* replied that testing kits were available for soil nitrate analysis in the U.S.A. and that many growers were using them.

Dr. D. B. Davies said that in advisory work in the south west he had found a disappointing number of blooms formed during the low light intensity period of the winter. He said it had been suggested that lower nitrate levels should be applied during the winter when the light was poor. *Dr. Davies* asked if the speaker had any information suggesting that the nitrogen should be reduced in the feed during the winter. *Dr. Winsor* replied that no experiments to date had been carried out on the strength of nitrogen feed for carnations during different times of the year but it was hoped to start such work shortly. However, he pointed out that during the winter less water would be required and normally less liquid feed would be given so less nitrogen would be applied. From his observations of other experiments *Dr. Winsor* thought that it would not be advisable to reduce nitrogen levels during the winter but possibly there was a case for increasing them.

Dr. J.K. R. Gasser commented on the striking difference between the nitrate contents of agricultural soils and those of glasshouse soils. Agricultural soils normally had less than 20 ppm of nitrate nitrogen whereas speakers at this meeting had mentioned levels in excess of 100 ppm in glasshouse soils. *Dr. Gasser* thought it would be wrong to try to equate agricultural and glasshouse practice because they had completely different sets of reference values.

Mr. G. A. Wadsworth asked *Dr. Winsor* what method of nitrogen analysis he would recommend for soil testing prior to planting. *Dr. Winsor* said that for many crops he would not make any determination for nitrogen because most glasshouse soils do not supply much nitrogen to the crop. He thought it might be helpful to estimate the soil for nitrate content before planting lettuce as excess nitrogen can be harmful but he did not think it would be necessary before planting tomatoes. The only exception to this would be for tomatoes planted in a very rich maiden soil which had been steam sterilized.

Dr. N. H. Pizer asked *Dr. Winsor* what experiments have been done to find out how applicable the results found at the Glasshouse Crops Research Institute were to glasshouses in other parts of the country. He wondered if the speaker had any empirical rules for applying results found at the Research Stations to glasshouses on very different soil types. *Dr. Pizer* also said that when the nitrate levels were increased the salinity levels in the soil were also increased and wondered whether any of the effects of increasing the nitrogen level might be due to increasing salinity. In answer to the first question *Dr. Winsor* said that there must be continual consultation between the Research Station and the advisers on application of their results in other glasshouses. With reference to nitrate levels and salinity *Dr. Winsor* said that nitrate was one of the most potent causes of salinity in glasshouse soils and excessive use of nitrogen could lead to disastrous results. Inevitably salinity levels increase as nitrate levels increase and there did not appear to be any way of avoiding this in experimental work. He thought that separating nutritional from salinity effects was a formidable and perhaps impossible task.

Mr. W. Dermott said he disagreed slightly with *Dr. Winsor* in what he said about the value of soil analysis for nitrogen prior to planting in glasshouse soils. He thought this could be very valuable when the planting was carried out in rich steam sterilized soils. *Dr. Winsor* agreed that this could be useful particularly in new glasshouse soils.

Dr. C. Bould commented that he thought it would be foolish to consider nitrate values of soil without considering also the phosphorus status of the plant. *Dr. Winsor* had mentioned this but he thought it should be re-emphasized.

Evaluation of nitrogenous fertilizers

By M. R. J. HOLMES, M.A., Dip.Agric. (Cantab.)
Levington Research Station

Introduction

EVALUATION of nitrogenous fertilizers, or indeed of any fertilizer materials, should provide information on their efficiency as suppliers of plant nutrients and their safety to crops under the normal conditions of use. Ways in which the efficiency of particular materials can be improved should also be investigated. Two phases, which are often in practice concurrent, are desirable. Examination under laboratory and glasshouse conditions can give a detailed assessment of the probable reaction of the fertilizer with the soil, the fate of the nitrogen, and the effect on crops. Field experiments are used to evaluate actual fertilizer behaviour in the field, usually by means of crop response and nutrient uptake. The results from the laboratory and glasshouse work can then be used as an aid in interpretation of the field results. In this review some consideration is given to methods of evaluating fertilizers in the field. The factors likely to influence the efficiency of nitrogenous fertilizers are discussed and the field evidence on the behaviour of nitrogenous fertilizers is reviewed with particular reference to conditions in Britain.

Methods of field evaluation

Evaluation of nitrogen sources in the field has to take account of two possibilities:

(i) The maximum response is the same for each source but different amounts of the various nitrogen sources are available to the crop.

(ii) The maximum responses differ because of toxicity or other effects.

Ideally, the sources under investigation should always be compared at a range of rates to see whether toxicity arises. However, this procedure can often limit the number of sources that can be compared, or the scope of the investigation, and some compromise is often necessary. Most comparisons of sources applied broadcast, to the seedbed or as top-dressings, will fall into the first category, and can justify making comparisons at one rate only; there will, however, be exceptions, exemplified by the scorch and yield depressions sometimes caused by high rates of gas liquor or calcium cyanamide applied as top-dressings to established crops and the reduced germination of small-seeded crops such as kale brought about by high rates of calcium nitrate or urea. Where placement of nitrogen sources is under consideration, or with sensitive crops such as tomatoes, toxicity is possible and should always be suspected in planning an investigation. Several rates of comparison are desirable, but where it is known that a certain material can damage the crop, and possible improved variants of it are to be investigated, comparison at a single high rate at which damage to the crop can be expected may well provide a satisfactory test.

If, in straightforward comparisons of availability, a single rate of comparison is used to provide information on the relative efficiencies of two or more nitrogen sources, it is important that this rate is on a sensitive part of the response curve. It can be shown that the optimum rate of test is dependent on the expected percentage difference between sources and on the shape of the response curve, but not on the size of the limiting response. The rates that should be used for the important crops under average conditions in Britain can be calculated from suitable response data. These rates are, as might be expected, well below the economic optima for arable crops, but for grassland are much higher than usual practice.

Interpretation of the data should take into account the non-linearity of response. Comparison of responses as such will over-estimate the value of the less satisfactory source, and the graphical method used by Cooke[8] and Cooke and Widdowson[10] to estimate percentage efficiencies is simple and adequate, provided the comparisons have been made on the rising part of the response curve. The mathematical methods available are laborious, and over-elaborate for the data under consideration.

With equal limiting responses, that is with relative efficiency unaffected by rate of application, calculations of values per effective unit of nitrogen need only take into account relative efficiency and cost per unit of nitrogen. With unequal limiting responses, however, economic evaluation is more difficult. It is necessary to take into account both the frequency and the size of possible harmful effects from an alternative source, which may well be ruled out completely for certain crops or methods of use.

Factors affecting the efficiency of nitrogenous fertilizers

The principal reasons underlying differences in behaviour of nitrogenous fertilizers are as follows:

Soil factors

1. *Leaching*. It is well established that forms of nitrogen differ in their resistance to leaching. Thus, Gasser[26,27], Cunningham and Cooke[13], and Cooke *et al*[11] demonstrated the more rapid leaching of nitrogen applied as calcium nitrate than as ammonium sulphate; ammonium nitrogen does not leach until nitrified. Urea, before conversion to ammonium, leaches less readily than nitrate[7]. Since it is rapidly ammonified in almost all soils its leaching propensities will normally be similar to those of nitrogen applied in the ammonium form.

The extent of leaching losses from different sources will depend chiefly on the amount of rainfall that percolates through the soil while the nitrogen is present as nitrate. Factors affecting this are:

(a) Excess of rainfall over evaporation and transpiration. Lower leaching losses in bare ground per inch of rainfall in summer compared with winter have been shown by Shaw[52].

(b) Presence of a crop and its rate of growth. Little nitrate is present in grassland soils[13,51], and an actively growing winter wheat crop can take up large quantities of nitrate in a period of a few weeks in spring[26,27]. On the other hand, Devine and Holmes[21] showed that nitrogen applied in early winter to grassland gave only small crop responses in spring, except with a very low winter rainfall, presumably because of leaching; ammonium, nitrate and urea forms of nitrogen all gave small responses. Ammonium nitrogen appears to nitrify slowly at normal winter temperatures in England and Wales, as was also shown by Gasser[26,27], and Shaw[52].

(c) Soil texture and structure. Shaw[52] found that nitrates were leached into the subsoil of two sandy soils, but disappeared completely after heavy rain from the sampled profile (to 24 in.) of two heavy soils. He suggested that either leaching through subsoil fissures or denitrification in the heavy soils accounts for its disappearance. Cunningham and Cooke[13] have suggested that nitrate can be held within the aggregates of heavy soils.

(d) Nitrification rate. Application some time before the crop will utilize the nitrogen can lead to greater leaching of nitrate sources than of ammonium sources. However, if nitrification is slow, as at low soil temperatures, leaching losses from ammonium forms will be less than from nitrate forms. In the U.S.A., autumn application of ammonium sources can be recommended in regions with a cold winter and low or moderate rainfall but not elsewhere[45].

The use of data from field experiments to draw conclusions on the prevalence of leaching losses from nitrates in farm practice is fraught with danger, because of the possibility that other influences such as volatilization or toxciity are operating. However, the data of Devine and Holmes[20,21], on winter wheat and grassland, Adams[1] on sugar beet, Nicholson and Hooper[47] on spring cabbage and Widdowson *et al*[67] on potatoes and kale, suggest that, at least in low and medium rainfall areas, leaching is not a cause of important differences between nitrogen sources. Results of further Levington experiments comparing spring applications of calcium nitrate, ammonium nitrate and ammonium sulphate in western England also suggest that leaching is not normally important (Table 1) as a cause of differences.

2. *Volatilization of ammonia.* Ammonium fertilizers and urea can be subject to loss of ammonia when applied on or near the soil surface[35,38,42,54,58,60,61]. Losses are greatest when the fertilizer is applied to the soil surface, and reduced by a soil cover. Volk's results show a large loss of ammonia from urea and a small loss from ammonium nitrate whether lime was added to the soil or not; ammonium sulphate lost a large amount of ammonia when the soil had been limed, and a small amount when it had not. Van Schreven, and Larsen and Gunary found that less ammonia was lost from ammonium nitrate than from ammonium sulphate applied to the surface of calcareous soils at equal ammonium rates. Larsen and Gunary's results show that the relative losses from ammonium nitrate, ammonium sulphate and diammonium phosphate, measured under laboratory and semi-field conditions, can vary with time. Losses tend to be greater under dry or nearly dry soil conditions, at high temperatures and on soils of low base exchange capacity or ammonia absorption capacity.

3. *Denitrification.* Loss of nitrogen as oxides of nitrogen or nitrogen gas may explain the considerable nitrogen deficits often found in nitrogen balance sheet investigations[3]. Shaw[52] suggests that it may have occurred in heavy soils after considerable rainfall; in this investigation approximately the same loss of of nitrogen occurred from the soil whether originally applied as ammonium sulphate or potassium nitrate.

4. *Fixation of ammonia.* Sohn and Peech[53] suggest that ammonia may be fixed more when applied as aqueous or anhydrous ammonia than when it is applied as a salt. Nommik and Nilsson[46] found that the amount of fixation was correlated with the organic carbon content of the soil. It is not clear how important this is as a cause of differences between nitrogen sources.

Table 1

The responses of barley, sugar beet, wheat and grassland to spring applications of calcium nitrate and ammonium nitrate compared with their responses to ammonium sulphate on acid and slightly calcareous soils in western districts of England

	Number of experiments	Mean responses (ammonium sulphate = 100)		S.E. of relative response	Number of experiments with a higher yield from CN than AS		Number of experiments with a lower yield from CN than AS	
		ammonium nitrate	calcium nitrate		Total	Significant ($P < 0{\cdot}05$)	Total	Significant ($P < 0{\cdot}05$)
Barley	1	109	131	24	1	0	0	0
Sugar beet:								
sugar	3	95	96	14	2	0	1	0
tops	3	103	96	12	1	1	2	0
Wheat	11	94	101	6	8	0	3	0
Grassland	11	104	107	4	7	1	4	0

Evaluation of bulky organic manures as sources of nitrogen

By J. WEBBER, B.Sc.

National Agricultural Advisory Service
Yorks. and Lancs. Region

LARGE numbers of experiments have been carried out in this country and elsewhere to measure the effects of bulky organic manures on crops but in only relatively few cases has the design been suitable for the evaluation of the manure in terms of plant nutrients. By far the biggest proportion of the published work relates to farmyard manure and only for this material has an assessment of its value in terms of nutrients been made for a range of crops. Other materials, poultry manure, sewage sludge, straw and various forms of compost have often been included in field experiments both long and short term but the data available on their supply of nutrients is usually of a general nature.

Farmyard manure

A big advance in our knowledge of the manurial requirements of crops was made by the publication of the classic paper 'The Fertilizer Requirements of Arable Crops' by Crowther and Yates[5] in 1941. They summarized the results of most of the trials carried out in this country in the period 1900 to 1940 and from a comparison of sites with and without farmyard manure calculated the average effect of the manure in supplying nutrients. On nitrogen they said 'the most remarkable feature of the results is the small difference between the responses to nitrogen in the absence of dung and in the presence of dung', and went on to suggest that the reason for this was the improvement in the physical condition of the soil and the fact that dung would be used on the more exhausted soils which together would then render the crops 'particularly responsive to the nutrients supplied by added fertilizers'. The exhaustion of nitrogen was likely to be greater than that of phosphate and potash and would account for the different results obtained. This finding, although unexpected, was generally accepted and was confirmed by later workers in their work with potatoes[6,18].

More recently Boyd[2] has examined all the available data relating to potatoes and has shown the importance of taking interactions between nutrients into consideration. Taking the results of 23 experiments carried out between 1937 and 1954 in which fertilizers were tested in the presence and absence of farmyard manure he showed that while the overall effect of nitrogen was slightly larger on plots receiving farmyard manure than where none had been used this was largely accounted for by the higher yields on plots getting no potash. In the absence of farmyard manure growth was restricted by potash deficiency so limiting nitrogen response while on the manured plots potash supply from the farmyard manure was adequate. Boyd went on to show that in the presence of phosphate and potash, farmyard manure reduced the need for fertilizer nitrogen. The following tables illustrate the importance of considering interactions between nutrients.

Table 1

Mean responses to nitrogen with and without farmyard manure and with different combinations of basal fertilizer

(mean of 23 experiments: tons per acre potatoes)

(after Boyd, 1959)

Basal fertilizer	O	P	K	PK	Mean
No FYM	0·89	1·13	1·92	3·12	1·77
FYM	1·61	2·01	2·48	2·09	2·05

Table 2

Effect of nutrient level on response to nitrogen

(mean of 13 experiments: tons per acre potatoes)

(after Reith and Inkson, 1958)

Basal fertilizer	P0K0	P1K1	P2K2	Mean*
No FYM	0·4	1·0	1·6	0·8
FYM	0·3	0·6	0·3	0·9

*mean response over all combinations of phosphate and potash.

Boyd[2] showed that in the different series of experiments considered by him, 10 tons per acre farmyard manure supplied from 0·15 to 0·35 cwt. per acre nitrogen with an average figure of 0·27 cwt per acre. There were, as might be expected, big variations between centres due to differences in the quality of the farmyard manure and to other factors.

Since the appearance of Boyd's paper a number of other estimates of the value of the nutrients in farmyard manure have been made for a range of crops. Adams[1] summarized the results of 41 trials on sugar beet, testing all combinations of three levels each of nitrogen, phosphate and potash in the presence and absence of farmyard manure, carried out in the period 1957 to 1960. In these experiments yields of sugar were reduced at higher levels of nitrogen and in the presence of 12 tons per acre farmyard manure the lowest nitrogen level (0·6 cwt/acre) gave the best results. Because of this no accurate estimate from yield of the value of the nitrogen supplied by the farmyard manure was possible but separate calculations were made based on sugar percentage, yield of tops and juice purity all of which are affected by the nitrogen but not by other nutrients. These showed twelve tons per acre farmyard manure to be equivalent to 0·3 to 0·6 cwt N per acre and Adams suggested reducing the level of nitrogen fertilizer for the crop by 0·4 cwt per acre (from 1·0 to 0·6) when farmyard manure is applied.

Patterson and Watson[17] compared 4 levels of farmyard manure on potatoes and sugar beet in combination with two levels each of nitrogen, phosphate and potash in experiments at Rothamsted and Woburn in 1956–57. The design of their trials was not suitable and the number too small for an independent estimate of the fertilizer equivalents of farmyard manure to be made. They did, however, test Boyd's conclusions and showed that their results confirmed his figure of 0·3 cwt N per acre from 10 tons per acre farmyard manure for potatoes. For sugar beet there were indications that farmyard manure supplied more nitrogen than for potatoes which supports the results obtained by Adams[1] described above.

A recent summary of a long term experiment carried out on the Leeds University farm by Holliday[12] suggests that although the effect of farmyard manure was greater in dry than in wet seasons the response to fertilizer nitrogen was also affected by rainfall, and on average 10 tons per acre farmyard manure supplied about 0·40 cwt N per acre. The following table giving maximum yields estimated from quadratic equations summarizes Holiday's results.

Table 3

Nitrogen dressings for maximum yield
(after Holliday, 1963)

	Nitrogen dressing—cwt N per acre 5 driest years	5 wettest years	Mean of 10 years
No FYM	1·52	1·41	1·44
FYM	1·07	1·02	1·04
	Maximum yield—tons per acre potatoes		
No FYM	12·22	12·57	12·37
FYM	13·66	12·89	13·08

A different approach was used by Williams *et al*[24] using micro-plots over the period 1956–60 at Rothamsted. Because only two levels of each nutrient were tested and there were only three plots with farmyard manure in each block accurate calculations based on yield were not possible. Estimates made on the basis of nutrient uptake, however, showed that about 0·27 cwt N per acre was supplied by 10 tons per acre farmyard manure both in the presence and absence of NPK fertilizer. Separate estimates for potatoes, kale and permanent grass showed that rather more nitrogen was supplied for potatoes and rather less for the other two crops.

In Scotland, Reith and Inkson[19] carried out experiments on swedes during the years 1948 to 1952. Dressings of 15 tons per acre farmyard manure were applied in 10 experiments with three levels each of nitrogen, phosphate and potash in all combinations. The farmyard manure used contained 0·37 per cent nitrogen, a lower value than that usually accepted (0·5 per cent), although in agreement with the average found by Hemingway[9] for 50 samples from the West of Scotland. Hemingway however determined nitrogen on dried samples so that his values might be expected to be rather lower than for determinations on fresh material. Reith and Inkson also state that their phosphate dressings were below optimum for swedes which could also have reduced the effectiveness of farmyard manure as a source of nitrogen. Different crops have different capacities for utilizing the nitrogen in farmyard manure[17] and the lower nitrogen requirement of swedes by comparison with some other crops could also be an important factor in this connection.

The following table summarizes the data on farmyard manure.

Table 4

Nitrogen equivalent of 10 tons/acre farmyard manure

Crop	cwt/acre N	Authority
Potatoes	0·30	Boyd
Potatoes	0·30	Patterson and Watson
Potatoes	0·40	Holliday
Potatoes, kale, grass	0·30	Williams *et al*
Sugar beet	0·40 (0·3—0·6)	Adams
Sugar beet	>0·30	Patterson and Watson
Swedes	0·12	Reith and Inkson

In view of the heterogeneous nature of farmyard manure and its varied composition as a result of different animal origin, proportion of straw or other litter and conditions of storage the agreement between the results of different workers is satisfactory and the now generally accepted value of 0·3 cwt N for 10 tons farmyard manure can be used with some confidence for most crops.

Slurry

The information on farmyard manure generally refers to cattle manure either from dairy cows or beef cattle for which straw was used as litter. Changes in farming practice have resulted in the use of water instead of straw and similar materials as an absorbent for the animal faeces and urine. The resulting slurry (or gülle as it is called in parts of Europe) could be considered as a bulky organic manure, but in view of its different physical form, will not be dealt with in detail. Experimental work has been reported by Herriott and Wells[10], Herriott *et al*[11] and McAllister[15] while further experiments have been carried out by the N.A.A.S. at Aberystwyth, Wolverhampton and Leeds. Variable results have been obtained, depending on the time of year when the material was applied and on the relative proportions of faecal and urine nitrogen present in the slurry. Winter dressings are inefficient and lead to considerable losses of nitrogen, but applications in spring indicate that the nitrogen in slurry may give between 50 and 80 per cent of the effect obtained from a similar amount applied as mineral fertilizer.

Poultry manure

Poultry manure, because of its higher nitrogen content, also needs to be considered separately. Tinsley and Nowakowski[21] investigated the manurial value of deep litter from poultry houses in two experiments on potatoes carried out in 1954–55. They used deep litter from laying hens produced over a period of 11 months in a house containing 250 birds in which straw, 22 cwt, had been used as litter. In a wet season, 1954, six tons per acre of the deep litter supplying about 2·0 cwt N per acre of which 42·5 per cent was soluble in cold 0·1 N sulphuric acid gave a yield increase similar to that obtained from 0·84 cwt N per acre as sulphate of ammonia. In this experiment the deep litter was approximately equivalent to fertilizer when compared in terms of 'soluble' nitrogen or about 40 per cent as effective in terms of total nitrogen. In the second experiment in 1955, using litter only five months old, the crop suffered badly from drought and there was no increase in yield from the use of deep litter although sulphate of ammonia gave an additional two tons per acre potatoes. The poor response in this year was attributed to the dry weather, the less mature compost and absence of basal nitrogen (0·35 cwt N per acre as sulphate of ammonia was given to all plots in 1954).

Further work as yet unpublished has been carried out on deep litters by Mudd at Great House Experimental Husbandry Farm in East Lancashire. Variable results have been obtained following the use of deep litter on grassland with the effectiveness of the litter in terms of nitrogen ranging from 50 to 100 per cent, based on the full production from the sward during the season of application. Brief details of this work have been given in the Great House Annual Reports.

Straw and straw composts

Straw and straw composts have often been advocated for improving the organic matter content of soils and their effect on nitrogen supply must be considered. Straw dressings tend to reduce the nitrogen available to crops, although this may later be slowly released. Many of our generally accepted ideas about the amount of nitrogen immobilized by straw come from some early work carried out by Hutchinson and Richards[13] at Rothamsted. They showed in laboratory studies and for composts that 100 parts of straw required 0·75 parts of nitrogen (0·15 cwt N per ton) and this amount has been recommended to farmers and used in experiments. It was found, however, in the three course rotation experiment at Rothamsted started in 1933 that yields following the use of straw at 53 cwt per acre with additional nitrogen were often higher than when the straw was composted with the same amount of nitrogen before application. The layout was revised and Patterson[16] has reported on the results obtained from 1953 to 1958. He showed that for straw ploughed in, it was sufficient to apply 0·08 cwt nitrogen per ton in order to compensate for the nitrogen absorbed from the soil during decomposition of the straw. When straw was composted, however, the larger amount suggested from the work of Hutchinson and Richards[13] was required while there was also loss of potassium from the straw by leaching during the composting process.

Harvey[8] described an experiment carried out at Sprowston comparing the effects of straw and farmyard manure on a six course rotation of potatoes, wheat, oats, sugar beet, wheat, oats. He found that the results from ploughing in straw were variable and depended on the summer rainfall. In wet seasons, straw gave better yields than farmyard manure and Harvey concluded that some of the extra nitrogen applied to help rot the straw was being used by the crop. Since extra nitrogen at the rate of 0·15 cwt N per ton of straw was used, this is in line with the conclusions of Patterson described above.

Straw compost has been tested in other experiments as it could be used in horticulture as a substitute for farmyard manure, and is generally more easy to handle than straw. Rothamsted experience suggests that there is no advantage in composting the straw, as both nitrogen and potassium are immobilized or lost by leaching. In the bulky organics experiments carried out on Experimental Horticulture Stations and reported by Webber[22,23] much higher rates of nitrogen (0·3 cwt N per ton of straw) were used in composting, when the compost generally gave rather better results than those obtained elsewhere. Variations in the composting process, especially in relation to nitrogen loss, are evidently important, but no published data seems to be available at present on the evaluation of the nitrogen in such compost.

Nitrogen immobilized by the rotting of straw in the soil or during composting does not appear to become available to crops within the first year or two after use, but Patterson[16] suggests that in the longer term residues from regular applications may be released as indicated by a rising trend in yields in the three-course rotation experiment. Laboratory work[3] and information obtained by cropping in a glasshouse[20] confirm that nitrogen in straw compost or immobilized in the soil when straw rots down is only slowly released. The effects noticed by Patterson[16] would of course come from the accumulated residues of applications of straw compost over a number of years.

Sewage sludge

The other organic manure which is available in large quantities and which could be used in agriculture and horticulture is sewage sludge. Some sewage sludges contain large amounts of metals such as zinc, copper, nickel and chromium which are toxic to plants and which make their use on the land undesirable, if not positively harmful. Other sludges, however, are less affected in this way and merit consideration. Bunting[4] has reported on the large series of trials testing sewage sludges both raw and digested and sewage sludge straw compost in comparison with farmyard manure and straw compost organized by Rothamsted during the years 1942 to 1949. The sludges used contained more nitrogen than the farmyard manure (0·9 and 1·4 per cent for the raw and digested sludges as compared with 0·6 per cent for farmyard manure) and at the rates used, 5 and 10 tons per acre dry matter, supplied more organic matter and nitrogen than 8 and 16 tons per acre farmyard manure. The average amounts of organic matter and nitrogen supplied were 3·3 to 3·8 tons per acre organic matter and 3·6 to 3·9 cwt per acre nitrogen for the sewage sludges, and 1·9 tons per acre organic matter and 1·4 cwt per acre nitrogen from the farmyard manure. The design of the experiments did not permit the calculation of nitrogen equivalents for the materials used, but the following table gives some of the results, with the interactions between the organic manures and fertilizer and nitrogen.

Table 5

Responses to organic manures and nitrogen
(tons per acre potatoes)
(after Bunting, 1963)

Manure and No. of experiments	Response to manure		Response to nitrogen		Interaction*
	No nitrogen	Nitrogen	No manure	Manure	
FYM (17)	2·67	2·43	1·79	1·55	−0·23 ± 0·24
Sewage sludge (20)	1·66	0·22	1·79	0·35	−1·44 ± 0·31
Straw sludge compost (9)	2·57	1·96	1·85	1·24	−0·62 ± 0·42

*Interactions are twice the interactions as formally defined by Yates[25] Nitrogen applied at 0·60 cwt per acre. Basal fertilizer 0·60 cwt per acre P_2O_5 and 1·20 cwt per acre K_2O.

Similar results were obtained on cabbage and on a number of horticultural crops. The large negative interaction between the sewage sludge and nitrogen and the very small effect of the sludge in the presence of fertilizer nitrogen shows that the sludge acted mainly as a source of nitrogen. The other materials used released smaller amounts of nitrogen but had other effects on crops, such as the supply of potassium and the improvement of physical condition (referred to by Bunting as 'more complex effects'). A rough estimate of the nitrogen supplying power of the sludges used suggests that 7½ tons per acre sludge dry matter (25 to 30 tons per acre sludge) supplied about 0·6 cwt N per acre for potatoes or 0·9 cwt N per acre for cabbage.

A further series of experiments on sewage sludge carried out during the same period in school gardens and reported by Garner[(7)] gave rather similar results. Farmyard manure was applied at 16 tons per acre and sewage sludges at 10 tons per acre dry matter with and without nitrogen at 0·6 cwt per acre. The following table summarizes some of the results obtained with potatoes on responsive sites.

Table 6

Responses to organic manures and nitrogen

(tons per acre potatoes)

(after Garner, 1962)

Manure and No. of experiments	Response to manure		Response to nitrogen		Interaction
	No nitrogen	Nitrogen	No manure	Manure	
FYM (17)	2·01	1·12	2·98	2·09	−0·89
Sewage sludges					
Digested domestic (8)	1·97	0·76	1·88	0·85	−1·03
Digested industrial (3)	2·19	1·02	2·51	1·34	−1·17
Raw (3)	3·54	−0·25	3·70	−0·09	−3·79

Generally, sewage sludge seems to be a good source of nitrogen as might be expected from the higher nitrogen content of the organic matter than in most of the other organic manures considered. In long term experiments it can give results approaching those obtained from farmyard manure, as shown in the bulky organic experiments on Experimental Horticulture Stations[(22,23)] but in other cases, possibly due to the presence of toxic metals, yields were little or no better than those on plots getting the same fertilizer but no organic manure.

References

1. Adams, S. N. (1962). *J. Agric. Sci.*, **58,** 219.
2. Boyd, D. A. (1959). *J. Agric. Sci.*, **52,** 384.
3. Bremner, J. M. and Shaw, K. (1957). *J. Sci. Fd. Agric.* **8,** 341.
4. Bunting, A. H. (1963). *J. Agric. Sci.*, **60,** 121.
5. Crowther, E. M. and Yates, F. (1941). *Emp. J. Exp. Agric.*, **34,** 77.
6. Edwards, G. H. A., Watkin, J. E. and Webber, J. (1956). *Expl Husb.*, **1,** 25.
7. Garner, H. V. (1962). *Emp. J. Exp. Agric.*, **30,** 295.
8. Harvey, P. N. (1959). *J. Roy. Agric. Soc.*, **120,** 55.
9. Hemingway, R. G. (1961). *Emp. J. Exp. Agric.*, **29,** 14.

10. HERRIOTT, J. B. D. and WELLS, D. A. (1962). *J. Brit. Grassl. Soc.*, **17**, 167.
11. HERRIOTT, J. B. D., WELLS, D. A. and CROOKS, P. (1963). *J. Brit. Grassl. Soc.*, **18**, 339.
12. HOLLIDAY, R. (1963). *Proc. 10th Easter School in Agric. Sci., Nottingham* 1963, 248.
13. HUTCHINSON, H. B. and RICHARDS, E. H. (1921). *J. Minist. Agric.*, **28**, 398.
14. MANN, H. H. and PATTERSON, H. D. (1963). *Rep. Rothamst. Exp. Sta.* 1962, 186.
15. MCALLISTER J. S. V. (1962). *Res. exp. Rec. Ministry Agric. N. Ireland*, **12**, 123.
16. PATTERSON, H. D. (1960). *J. Agric. Sci.* **52**, 222.
17. PATTERSON, H. D. and WATSON, D. J. (1960). *Rep. Rothamst. Exp. Sta.* 1959, 164.
18. REITH, J. W. S. and INKSON, R. H. E. (1958). *J. Agric. Sci.*, **51**, 218.
19. REITH, J. W. S. and INKSON, R. H. E. (1963). *J. Agric. Sci.*, **60**, 145.
20. STEWART, B. A., JOHNSON, D. D. and PORTER, L. K. (1963). *Soil Sci. Soc. Amer. Proc.*, **27**, 656.
21. TINSLEY, J. and NOWAKOWSKI, T. Z. (1959). *J. Sci. Fd. Agric.*, **10**, 232.
22. WEBBER, J. (1961). *Expl. Hort.* **5**, 53.
23. WEBBER, J. (1963). *Expl. Hort.* **9**, 39.
24. WILLIAMS, R. J. B., COOKE, G. W. and WIDDOWSON, F. V. (1963). *J. Agric. Sci.*, **60**, 353.
25. YATES, F. (1937). Tech. Comm. 35, *Comm. Bur. Soil Sci. Harpenden.*

Discussion

Mr. M. R. J. Holmes asked *Mr. Webber* if he thought farmyard manure would have as much value for early potatoes as for main crop potatoes. *Mr. Webber* replied that he did not know of any experiments testing farmyard manure on early potatoes but he would guess that it would be less effective on early potatoes than on main crop potatoes because the former were in the ground for such a short time.

Dr. T. Z. Nowakowski said that the main constituent containing nitrogen in poultry manure was uric acid and the best way to conserve this nitrogen was to dry the manure rapidly at 100°C. If this is not done the uric acid changes rapidly to ammonia which could be lost. *Mr. Webber* replied that he agreed this would be the best way to dispose of poultry manure from batteries but the cost of drying poultry manure was too high to be economic.

Mr. W. Dermott commented that recent published figure for the cost of drying poultry manure was £17 per ton and at this price he thought the drying process was uneconomic.

Dr. C. Bould commented on the effects of plant viruses on the nitrogen nutrition of crops. He said that when strawberries were infected with viruses, large applications of fertilizer particularly nitrogen were needed to get a satisfactory crop. Since viruses have been eliminated from the strawberry plant it has now been found that the strawberry has a very low nitrogen requirement although the internal cell requirement is much the same. When viruses are eliminated from other fruit crops they also may have a much lower nitrogen requirement. *Dr. Bould* also said that he thought too much attention was being paid to field experiments which simply define what happens under certain conditions but make no fundamental contribution to knowledge. He thought that more fundamental work was necessary on soil plant relationships and on the actual nitrogen requirement of the crop itself.

Nitrogen and advisory problems

By W. DERMOTT*, M.Sc.
National Agricultural Advisory Service
West Midland Region

IN the course of his work the adviser has to deal with a large number of problems concerning soil nitrogen and the use of nitrogenous fertilizers. Expenditure on nitrogenous fertilizers in this country, currently about £50 million per year inclusive of subsidy, is increasing steadily and so far as can be foreseen is likely to continue to increase, particularly for use on grassland. Nitrogenous fertilizers usually have marked effects—generally beneficial but occasionally harmful—on the growth of crops and grass and farmers are realizing more than ever that the beneficial effects can, with proper management, give rise to greater profits.

It is the purpose of this paper to indicate some of the problems which the adviser encounters in a range of crops and in different soils and situations. Some stem from insufficient acceptance by the farmer of existing knowledge but it can be expected that these will be rectified in time, either by the continuing process of aducation and advice or by economic pressures or both. In other instances, however, there is not enough knowledge at present to give sound advice. Where examples are quoted they are taken from the West Midland Region. This is simply a matter of convenience so far as the writer is concerned and in some cases more striking figures could be quoted for other parts of the country.

Soil nitrogen

The prediction of the likely response to nitrogenous fertilizers by the estimation of some fraction of soil nitrogen or an associated factor could be a useful advisory tool and this is discussed further in a later section.

There are several other problems concerned with soil nitrogen on which further information is needed. One of these concerns the relationship between nitrogen and soil structure. It is a matter of common observation that crops and grass on poorly structured soils often look nitrogen deficient and need abnormally heavy dressings of nitrogenous fertilizers. Is this, despite the mobility of nitrates in the soil, because the sparse root systems often found under such conditions are inefficient extractors of nitrogen or are the nitrogenous fertilizers suffering large denitrification losses?

The relationships between nitrogen and water are of great interest. There is a certain amount of experimental backing for the observation that nitrogen can replace water to some extent and help plants to grow under dry conditions. With irrigation, is it correct advice to recommend heavier than ordinary dressings of nitrogenous fertilizers because large crops are grown or will normal dressings suffice because they are used more efficiently?

Perhaps the most pressing thing the adviser needs to do, is to study the soil and other factors which lead to unusually large and unusually small responses to nitrogenous fertilizers. Also, further study is required of the con-

*Now stationed at N.A.A.S. Sub-centre at Wye.

ditions under which abnormally large losses of nitrogen occur because of denitrification or for other reasons. Early spring top-dressing of nitrogenous fertilizers sometimes seem to be very inefficient and fertilizers applied during drought may have little effect when rain comes.

A minor but interesting problem is the reaction of nitrogenous fertilizers with the strongly alkaline material, pulverized fuel ash. The increasing production of pulverized fuel ash will bring this question into greater prominence.

Prediction of nitrogen response

Much will have been heard at this conference about the prediction of nitrogen response based on a variety of chemical methods and incubation techniques. Whether such tests can give valid results under all conditions is very doubtful any more than a method for 'available' phosphorus or potassium can be universally applicable. There are many reasons for this including the possibility of there being nitrogen reserves at depth in the soil, the effects of structure and moisture supply, the unpredictability of future weather, etc.

Notwithstanding such limitations, soil tests for prediction of nitrogen response are likely to be of considerable value to the adviser both in dealing with specific problems and also in characterizing different soils, and soils in various cropping systems, in relation to their capacities for supplying nitrogen to plants. At the present time it seems unlikely that such tests can be carried out on large numbers of samples because of the work involved in taking such samples and dealing with them afterwards in the laboratory.

Tests for the prediction of nitrogen response, of course, only give information about soil status and have to be linked with field experimental results and experience before they can be used to give advice on the use of nitrogenous fertilizers. In this connection, it should be borne in mind that as new varieties of crops are introduced there is likely to be a slow change in optimum rates of nitrogenous fertilization.

The combination of soil test, field experimental evidence and experience has been employed to construct the table used to give advice on spring nitrogen for cereals in the West Midlands, which is reproduced in the appendix to this paper. Whether soil tests on individual fields would enable better advice to be given would make a subject of useful study.

Related to the subject of the prediction of nitrogen response, there is a need to collect information on responses of crops on different soil series and soil types, e.g., in the West Midlands, potatoes grown on the Evesham Series (a heavy textured soil derived from Lias Clay) respond poorly to nitrogen whereas much larger responses are obtained on the Newport Series[3] (a light textured soil derived from Triassic drift). Although this sort of information is valuable it needs to be supplemented by an investigation of the factors which cause such differences.

Choice of nitrogenous fertilizer

This is a subject of considerable interest to the adviser as the number of nitrogenous fertilizers increases. These now include sulphate of ammonia, the ammonia nitrate/lime fertilizers, ammonium nitrate and urea. In practice, urea is being used on an increasing scale because of its low cost per unit of nitrogen, high concentration which reduces handling costs, suitability for aerial application, etc. Unfortunately it would seem that urea is likely to be somewhat less efficient, and occasionally much less efficient, than other nitrogenous fertilizers. Probably we are not using urea in ways calculated to get the best out of it; for example, should urea and other nitrogenous fertilizers be cultivated into the soil more deeply by being applied at an earlier stage of seedbed preparation, than is the practice on many farms today. Such measures should help to keep ammonia losses to a minimum.

Nitrate fertilizers, e.g. sodium and potassium nitrates, have a high efficiency for many purposes and would be very useful in British agriculture if they were comparable in cost with other nitrogen sources.

Liquid fertilizers are being used on a small scale in this country and their use may well increase, not because they necessarily have any technical superiority over solid fertilizers, but they are easy to handle by suitably equipped farmers and contractors. Injection of liquid fertilizers into the soil could be expected to increase the efficiency of the nitrogen which is often derived wholly or in part from urea. Anhydrous ammonia, ammonia solutions and the like although widely used in the U.S.A. and elsewhere are not yet used commercially here. However, they may well have their place in districts of large-scale arable farming.

Gas liquor can be a satisfactory source of nitrogen, particularly for applying to arable land some time before sowing crops, in areas such as the West Midlands where it is cheap. If it is used under conditions when scorching of growing crops or grassland occurs, however, its efficiency is much reduced. Occasionally its scorching properties are made use of deliberately, to give a partial measure of weed control in kale crops.

In compound fertilizers today, nitrogen can be derived from several fertilizer sources and sometimes the adviser is at a disadvantage because he does not know what these sources are.

Increasing concentration of fertilizers calls for improved efficiency in spreading. Admittedly there is very little experimental evidence on the possible losses from poor spreading but the visual evidence on some farms suggests that better spreading is necessary.

'Slow-release' fertilizers such as ureaformaldehyde polymers, metal ammonium phosphates, coated fertilizers and substances for slowing down nitrification could be of value, particularly in horticulture, if they were sufficiently reliable and sufficiently cheap. Limited experience with some of them, however, suggests that they are not yet developed to the stage where their use can be recommended with any confidence except possibly for one or two specialized applications such as potting composts.

Organic sources of nitrogen

At the same time as farmers are using increasing amounts of nitrogenous fertilizers, an even larger amount of nitrogen in farmyard manure, poultry manure, slurry, liquid manure and sewage is being used inefficiently. All these materials have to be disposed of in some way and taking factors such as cost and avoidance of river pollution into account, application to the land is likely to be the best long-term solution. Admittedly with sewage there is often the problem of toxic metals but if this can be overcome then sewage is a useful source of nitrogen.

The role of the adviser should be to ensure that, within reasonable economic limits, organic manures are conserved and used to the best advantage. Unfortunately this is far from being the case on most farms. Only rarely, for example, is there any reduction in use of nitrogenous fertilizer when farmyard manure is used and yet this is a simple example of circumstances under which a worthwhile saving in fertilizer costs could be made.

Arable crops

In the arable districts, nitrogenous fertilizers are now being used at heavy rates, often in excess of the known requirements as determined by experiment. In East Shropshire, for example, in 1962 main crop potatoes received on average 140 units/acre of nitrogen and sugar beet 132 units/acre[2]; 48 per cent of the potato acreage and 28 per cent of the sugar beet acreage received farmyard manure in addition at an estimated rate of 13 tons/acre. Experiments in the same district or on similar soils elsewhere[1,3,4] suggest that the average optimum for both crops is around 100 units/acre in the absence of farmyard manure.

This is a situation which the adviser cannot view with any satisfaction and it is interesting to speculate how it has arisen. Mainly it seems to stem from three factors—the low cost of fertilizers relative to other farming costs, the understandable desire of farmers to simplify management by using a limited range of fertilizers and application rates and the view that heavy dressings of fertilizer will ensure a good crop. In addition advisers, who have their fair share of scientific caution, sometimes fail to give advice in a manner which carries conviction.

The figures quoted above are averages, of course, and conceal wide variations in practice from farm to farm. Nevertheless there is little evidence that farmers make much allowance for the factors which are known to influence nitrogen response, e.g., soil type, previous cropping and manuring, use of organic manures, winter rainfall, etc.

By contrast, in the mainly grassland districts crops, and particularly fodder crops such as kale and rape, are often given too little nitrogen. It is true that farmers sometimes hope that kale and rape will stand well into the winter and because of this do not wish to have a lush crop. Nevertheless there are many instances where more nitrogen could be given with advantage.

Cereals present a number of special problems. The average figures for nitrogen consumption often look reasonable in relation to the system of farming in particular districts but an examination of individual farm practice often shows no rational connection with previous cropping and manuring. The example from East Shropshire[2] in Table 1 shows an extraordinarily large usage of nitrogen on winter wheat grown after sugar beet.

Table 1

Use of nitrogenous fertilizer on winter wheat by previous cropping

Previous crop	Percentage of acreage receiving N	Average actual rate units/acre of N
Cereals	94	56
Potatoes	89	58
Sugar Beet	87	75
Grassland (generally short term leys)	95	53

Again, although there is a considerable body of information on optimum time of spring top-dressing of winter cereals in relation to soil nitrogen level and previous cropping, this is little used in practice. Time of top-dressing tends to be something which is determined on a farm basis, the time of top-dressing on any particular farm being adhered to fairly rigidly, rather than on a field basis. Many farmers, perhaps the majority, also give some nitrogen at the time of sowing winter cereals and although this can sometimes be useful, on the whole it is an inefficient practice.

The question of whether nitrogen combine-drilled is more efficient than nitrogen broadcast for cereals is, perhaps, still not entirely clear, despite a certain amount of recent work. Further information on this point is desirable because it has some bearing on the topic often raised by farmers nowadays, of whether or not to use a combine drill.

It is often found that farmers give considerably more nitrogen to wheat than to barley; in East Shropshire(2) wheat (winter and spring) received on average about twice as much nitrogen as spring barley, the latter being used largely for feeding. Although it is reasonable that wheat should receive rather more nitrogen than feeding barley, there are also some grounds for thinking that farmers have not yet fully realized the responsiveness of barley to nitrogenous fertilizers.

As with potatoes and other crops the responsiveness of cereals to nitrogen on different soil series is of interest. In 1964 yields of cereals were measured in seven fields each of which contained the light textured Newport Series and the heavier textured Salop Series. In this season, which was a dry one, the latter out-yielded the former by about 3 cwt/acre. An obvious extension of such a study will be to see if extra nitrogenous fertilizer will increase the yield of the Newport Series nearer to that obtained from the Salop Series.

With sugar beet the question of sugar content is important in addition to gross yield of roots and it has already been pointed out that many growers are giving nitrogenous fertilizers in excess of the experimentally determined requirements. With increasing emphasis on sugar percentage there is even more need to have a critical look at sugar beet manuring.

A more general but very important problem concerns the lower efficiency of utilization of nitrogenous fertilizers by arable crops than by grassland. One can only speculate as to why this should be so—is it for example because, at any rate in the early stages of growth, the roots of arable crops explore the soil much less thoroughly than the roots of grasses? If so, will close drilling or broadcasting of cereals or different spatial arrangements of potatoes made possible by the use of herbicides increase the efficiency of nitrogen use? The limited evidence available hardly suggests that this is so but further experimental work is clearly desirable.

Grassland

Entirely different considerations apply to grassland. With arable crops there is a tendency to use too much nitrogen whereas on grassland despite some improvement in recent years, there is still great scope for increased applications. Broadly, three systems of nitrogen usage have evolved—high usage (say 200 units/acre or more per annum) which is practised on a small but increasing number of farms, medium usage (say 100 units/acre per annum) practised on a large number of dairy farms and low usage (say 30 units/acre or none at all) where the main reliance is on clover nitrogen. Which system is adopted depends more on the degree of intensification required than on any technical consideration. It is to be expected that the number of farmers on medium and high nitrogen systems will increase and indeed there is already a considerable amount of interest in the possible economic and management problems associated with using nitrogen in the 300–400 units/acre range. Increasing use of nitrogen does, of course, raise certain technical problems and notably that of the possibility of outbreaks of 'metabolic' diseases of animals such as hypomagnesaemia. 'Nitrate poisoning', which can also occur on fodder crops such as kale and rape, is another possible hazard. Copper and cobalt deficiencies also can be associated with grassland improvement although not perhaps very directly with nitrogen usage as such.

The possibility of metabolic disorders occurring has not in general deterred advisers from recommending high nitrogen systems on gassland or farmers from adopting them. It is merely a factor to bear in mind so that suitable precautionary measures can be taken.

A difficult problem on farms which use large amounts of nitrogen on grassland is to decide suitable rates of application of potassic fertilizer particularly under grazing conditions. On heavier soils of satisfactory potassium-supplying power some farmers appear to be obtaining good results following the application of heavy dressings of nitrogenous fertilizers together with very small amounts of potassium. Whether this policy is satisfactory in the long term is a subject on which we need further information.

The growing realization that at high levels of manuring the production from permanent grass can approach that from leys may well revolutionize our ideas about grassland management and manuring, particularly on the heavier soils where leys can be difficult to establish. An example can be quoted from a grassland survey carried out in Cheshire in 1961–62[5] where at 140–160 units/acre of nitrogen per annum, virtually the same production of Utilized Starch Equivalent was obtained from both permanent grass and leys.

Although the general question of the level of nitrogen use on grassland is tied up with a number of rather complicated factors related to economics and the farmer's preference, there are some more straightforward cases where extra nitrogen would clearly be an advantage. The hay crop, for example, is often under-manured and paradoxically less nitrogen is used on hay in grassland than in arable areas, yet it is in the former that there is often a shortage of hay. In the predominantly grassland district of North Herefordshire the average use of nitrogen on permanent grassland cut for hay is 20 units/acre (10 per cent of the acreage received farmyard manure in addition), whereas in East Shropshire the equivalent figures are 32 units plus 25 per cent of the acreage receiving farmyard manure[2]. More extreme examples than this could, of course, be quoted for other parts of the country. Even making due allowance for the fact that heavy crops of hay are difficult to 'make' in high rainfall districts, there would seem to be a clear case for increased nitrogen usage on fields cut for hay.

On farms using modest amounts of nitrogen for grazing there is a need to examine times of application. A very high proportion is given in early spring and although extra grass is valuable at this time of year extra grass in midsummer or late autumn can often be useful also.

Nitrogen usage in the uplands and hills is at present negligibly low. With the increasing pressures on land in the lowlands it can be expected that there will be greater interest in hill land improvement and with it some increase in the use of nitrogenous fertilizers.

Vegetables

With vegetables, as with most horticultural crops, the advisor is handicapped by inadequate experimental evidence on the manuring of individual crops under a range of soil conditions. It is not suggested that experiments give answers to all the questions raised in practice but they do provide a background on to which the adviser can graft his experience. In recent years, for example, some hundreds of experiments have been carried out on the nitrogenous manuring of cereals, but with horticultural crops there are often only one or two experiments to fall back on, or none at all. This situation can be expected to right itself slowly but in the meantime insufficient experimental work remains the biggest advisory problem.

There is little recent Survey of Fertilizer Practice data on the use of nitrogenous fertilizers on vegetables but experience in some of the main vegetable districts suggests that the growers have now realized the value of nitrogen for certain crops and are using heavier rates than formerly. On occasions it is probable that too much nitrogen is given; in the Evesham district of Worcestershire, for example, 87 per cent of the green pea acreage received nitrogen at an average rate of 81 units/acre[6].

The relationship between nitrogen supply and 'quality' in vegetable crops needs further study. It is known, for example, that internal browning of Brussels Sprouts can be influenced by nitrogen level but further information is required in this connection on factors such as nitrogen/potassium interactions. Also rots of various kinds (Botrytis, 'soft rots', etc.) can be a cause of great loss in vegetables and we need to know more about the effect of manuring in predisposing crops to these conditions.

Glasshouse crops

A number of problems involving nitrogen occur under glass. One of these concerns seed and potting composts using sterilized loam where not infrequently seedlings of susceptible crops may grow poorly or die. Manganese toxicity is known to cause one type of injury but there is another which has long been known to advisers as 'ammonia damage'. There is, in fact, no evidence that ammonia is responsible and the name has been given simply because the composts contain abnormally high levels of ammoniacal nitrogen. It is possible that nitrogenous compounds are not involved at all. However, there is a problem here which needs to be cleared up and perhaps both chemists and bacteriologists are involved in this.

Prediction of nitrogen response could be of help in glasshouse advisory work because of the large release of mineralized nitrogen which can follow steam sterilization and the use of some chemical sterilants. If it were known that certain soils would release considerable amounts of nitrogen then suitable steps could be taken in the base manuring programme to correct the situation.

Determination of nitrate levels can be of value in two ways. Firstly they can indicate when nitrates are too low for optimum crop growth—although here we are handicapped by not having sufficient standard data on crop requirements—and secondly it is possible to detect cases when nitrates are likely to be at a damagingly high level. Except in soils containing large amounts of unusual anions such as chlorides, high levels of nitrates appear to be the main cause of salinity problems in glasshouse soils in this country.

So far as the individual crops themselves are concerned there is a considerable body of evidence on the nitrogen requirements of tomatoes but much less on any of the other glasshouse subjects.

Fruit

Here again, with the possible exception of the strawberry, there has only been a limited amount of experimental work on the nitrogen manuring of fruit. Perhaps there is reason to believe that some growers use too much. With increasing emphasis on fruit size and fruit quality much more information is required. Also there are some special points concerning nitrogen—the effect of urea sprays on fruit set, the use of post-harvest urea sprays and the optimum time of application of solid fertilizers. All these subjects are under examination by the Research Stations and the N.A.A.S. but the work is hardly far enough advanced as yet to enable firm recommendations to be made. Certainly some very good crops of fruit are being produced by growers who apply solid nitrogenous fertilizers in early spring in the conventional way.

There has been much discussion recently on the value of leaf analysis as a guide to the manuring of fruit crops and certainly the adviser has found leaf analysis a useful technique in problem cases for many years. Whether it is sufficiently reliable under a wide range of soil and crop conditions to enable much better advice to be given than that which is currently given based on experience plus limited experimental evidence is still an open question. Leaf analysis has been mentioned here with specific reference to fruit crops but, of course, similar arguments could be advanced in respect of other crops.

Hops

Hops are like many other horticultural crops in that a practice of nitrogenous manuring has evolved based on experience and some experiment which generally gives satisfactory results. Points on which further information is required and which are either under investigation or on which investigations are proposed include the rate and timing of nitrogenous fertilizers for hops grown in grass, the use of nitrogenous fertilizers in conjunction with modern systems of soil management such as mulches and the use of herbicides and the special value, if any, of concentrated organic sources of nitrogen to the hop.

Summary

Better application of existing knowledge is required on points such as the following:

(a) Optimum rates of application of nitrogenous fertilizers for crops such as potatoes and sugar beet. At present excessive dressings are often given.

(b) The relationship between rate and factors such as previous cropping and manuring.

(c) The factors influencing time of top-dressing of cereals.

(d) The need for higher rates of application of nitrogen to hay and kale and rape particularly in the grassland districts.

(e) The desirability of making better use of organic manures as sources of nitrogen and other nutrients.

Further information is required on:

(a) The relationship between nitrogen and soil structure.

(b) The relationship between nitrogen and water.

(c) The responsiveness of soil series and soil types to nitrogenous fertilizers and the factors which lead to variations in response.

(d) The value of soil tests and leaf analysis for predicting nitrogen response.

(e) Methods of obtaining higher efficiency from fertilizers such as urea.

(f) Methods of improving the efficiency of utilization of nitrogen by arable crops.

(g) Suitable rates of application of potassic fertilizers to 'balance' nitrogen dressings on grazed grassland.

(h) The nitrogen requirements of many horticultural crops in relation to yield and quality.

References

1. ADAMS, S. N. (1962). *J. Agric. Sci.* **58,** 219.
2. BLOOD, J. W. and others. 'Survey of Fertilizer Practice, 1962' (Preliminary report). Mimeographed report by the Ministry of Agriculture, Fisheries and Food.
3. BOYD, D. A. and DERMOTT, W. (1964). *J. Agric. Sci.*, **63,** 249.
4. EDWARDS, G. H. A., WATKIN, J. E. and WEBBER, J. (1956). *Experimental Husbandry* No. 1.
5. HOPE, R. (1964). *Agriculture (Lond.)*, **71,** No. 2. 68.
6. PHILLIPS, W. L. H., ENGLEDOW, R. A. and HILLS, MARY G. 'Fertilizer Usage on Market Gardens in the Evesham-Pershore District of Worcestershire, 1957'. Duplicated report issued by the Ministry of Agriculture, Fisheries and Food.

Appendix

Spring Nitrogen for Cereals in the West Midlands

The nitrogen requirements of cereals are affected to some extent by season and type of soil but the suggested rates in the table below should provide a guide to optimum levels of nitrogen manuring.

Cropping system	Probable soil nitrogen status	Units of nitrogen per acre				
		Winter wheat	Winter oats	Spring wheat	Spring barley (feeding)**	Spring oats
1. Arable or short ley-arable rotations. Cereals after arable crops or short term leys.	Moderate or low	60*	40*	60	60	40
2. Cereals after 3 year leys, red clover leys, arable crops receiving F.Y.M.† or root crops grazed off by livestock. Third or fourth crops after permanent grass.	Moderately high	45	30	40	40	20
3. Cereals after long leys or as 1st or 2nd crops after permanent grass. Fields receiving regular heavy dressings of F.Y.M. or poultry manure.	High	30	20	20	20	Nil

*On old arable soils not receiving F.Y.M. 70 or 80 units may be justified for winter wheat and 50 units for winter oats. Heavy dressings may also be required if 'Take All' is present.

**Malting barley should receive about half the nitrogen recommended for feeding barley.

†By F.Y.M. is meant a heavy application—15 tons or more—of reasonably good quality.

When nitrogen is applied to winter cereals in the autumn, the spring top-dressing should be reduced by half the amount applied in the autumn.

Top-dressings for winter cereals should be applied in March on soils of low and moderate nitrogen status (1 above) and in early May on soils of moderately high or high nitrogen status (2 and 3 above).

When fields of winter cereals are grazed the top-dressing should be applied soon after grazing is finished.

Nitrogen for spring cereals is best applied in the seedbed. It may be advantageous under some conditions to apply it late or split the application but the extra work involved is unlikely to be worthwhile.

Reports of Syndicates

THE Conference was then divided into Syndicates which were asked to consider specific questions and to prepare reports on them.

Question 1

What are the advantages and disadvantages of the various laboratory methods in assessing soil nitrogen status and nitrogen fertilizer recommendations?

Report—Syndicate 1*A*

Chairman: Mr. R. D. Russell

Spokesman: Mr. C. Berryman

The task of the adviser is to give positive advice. When giving advice on nitrogen use there are many unknown factors and the Syndicate has, therefore, assumed that the N.A.A.S. Soil Scientists require methods of analysis to be used in addition to the methods of prediction at present used. In looking at such methods consideration was only given to papers presented to the Conference.

Advisers have to consider the accuracy and suitability of a method as an agronomic tool for advice and also the adaptability of the method for routine use in a laboratory.

The Syndicate considered mainly field crops, but recognized that there are problems in glasshouse soils; in particular, methods for forecasting the effects of steaming on soil nitrogen need to be investigated. The Syndicate was unable at present to make any comments on suitable methods for this work.

Perennial fruit crops present special problems because reserves are present in the woody tissues, and the roots are exploring and re-exploring the same soil. Leaf analysis despite sampling difficulties in the field has possibilities for giving advice on the nitrogen manuring of perennial fruit.

Glasshouse soils may need sampling techniques linked to the method of feeding. The Spurway method merits a further investigation because of the advantage of its simplicity.

In any further investigations the effects of depth of ploughing should be investigated.

The biological methods have the inherent disadvantage of requiring special precautions in dealing with the soil sample.

The following methods* were not considered worthy of further investigation:

Method 2. Aerobic incubation at 25 °C for 28 days at 60 per cent water holding capacity.

The method is tedious and time-consuming for routine work. It correlates well with other methods and is thus useful for research work.

Method 3 (*Waring and Bremner*, 1964). The disadvantage of this method is

*Appendix to discussion on 'Determination of nitrogen requirement of crops by analysis' (see page 95).

that the period of incubation is 16 days and this has no advantage over Method 4 where the incubation time is seven days.

Method 6 (*Purvis and Leo*, 1961). This method is analytically difficult, and correlated poorly with the other methods.

Method 8 (*D. S. Jenkinson*, 1964). This method would have to be adapted considerably for routine work. Analytical problems can arise when this method is used: for example the possible deterioration of the barium hydroxide and contamination of cellulose from filter paper.

Aspergillus Niger. This method is lengthy and time-consuming and has no advantages over incubation methods. Similar remarks apply to measuring the carbon dioxide produced from cellulose using nitrogen available from the soil.

Leaf analysis. This method has no advantages for annual crops.

The following methods were considered worthy of further investigation:

Method 1 (*D. J. Eagle*, 1961). This method has the advantage that water holding capacity does not have to be determined and only seven days is required for incubation. The main disadvantage for routine use is that the sample must be incubated within two days of air drying. The critical nature of the two-day period after air drying, the temperature of the air drying and the possible effect of the vermiculite fixing ammonia, needs to be investigated if this method is to be considered for advisory purposes.

Method 4 (*Waring and Bremner*, 1964). This is the simplest of all the biological methods. It is straightforward and could easily be used on a routine basis. The main disadvantage is the seven days required for incubation.

Method 5 (*Truog*, 1959). This method has the advantage of being a chemical method which is rapid and can easily be adapted for routine work. The disadvantage is it does not differentiate soils giving a medium response from those giving a high response. However, the method is capable of clearly separating those soils where a low response is to be expected.

Method 7 (*A. H. Cornfield*, 1960). This method has the advantage that it is fairly short, requiring only two days to obtain the result. It is simple analytically, but difficulties arise with re-wetting peat soils. It differentiates low from high or medium responses but does not differentiate medium from high responses.

Inorganic nitrogen in potassium sulphate extract of fresh soil. This is ideal for special circumstances, for example giving advice on cereal top dressing in the spring and differentiating between good and poor areas in a field. The method is simple, but an inherent disadvantage is that of transporting fresh soils to the laboratory because the determination must be carried out within 48 hours.

Report—Syndicate 1*B*

Chairman: Mr. B. Wilkinson

Spokesman: Mr. M. F. Harrod

The problem was dealt with from an advisory context where the methods must be simple, reliable and suitable for normal routine analytical organization.

The syndicate considered the soil as a three-dimentional body where factors other than mineral nitrogen in the surface 6 in. were of significant importance.

It is important to have the soil series characterized on a profile basis to provide information on factors likely to influence nitrogen availability. This would lead to an equation in which 'available nitrogen' is only one of the factors. A soil might be characterized for the release of nitrogen by an equation from such data as:

(a) Active and passive organic carbon.

(b) Carbon/nitrogen ratio.

(c) Ammonia fixation capacity.

(d) Capacity of subsoil to store mineral nitrogen.

This equation would refer to mineral soils only, as peat soils must be considered as a special case and treated separately.

Methods of estimating 'available nitrogen' are divided into the Biological and chemical methods. They were looked at from two aspects:

(a) Analytical convenience.

(b) Value for giving fertilizer recommendations.

Total nitrogen estimation

This is of little value for direct advisory work but is a useful factor in characterizing soil series and soil profiles.

*Biological methods**

It was felt that these methods needed more standardization of technique and developments to this end are being pursued.

Methods 1 *and* 2—*aerobic incubation.* Both methods are somewhat tedious. In No. 2 the determination of water capacity is laborious, whilst in No. 1 some ammonia may be fixed by the vermiculite.

Methods 3 *and* 4—*anaerobic incubation.* These methods were found to be analytically very convenient due to the direct distillation after incubation. A disadvantage is the incomplete estimation of nitrate already present in the soil; this may be serious for advisory work on glasshouse crops.

Methods 5, 6, 7 *and* 8—*Chemical methods*

The techniques of Truog (No. 5), Purvis and Leo (No. 6) and Cornfield (No. 7) were tested and were felt to be inferior to the biological methods, especially No. 6 which gave poor correlations. The method of Jenkinson (No. 8) was felt to be a new approach as it aims at estimating fractions of organic carbon in the soil which govern the activity of micro-organisms which in turn determine the rate of nitrogen release. The disadvantages at present are:

(a) Very low levels of extraction

(b) Working with concentrated sulphuric acid

(c) The method has not been tested on a wide range of acid soils, nor on soils where nitrogen is immobilized, due to the recent incorporation of fresh organic matter.

Further development of this method is necessary to improve its efficiency and laboratory convenience.

*Appendix to discussion on 'Determination of nitrogen requirement of crops by analysis' (see page 95).

Summary

It was agreed that 'available nitrogen' data, however determined, must be accumulated on a logical basis. The interpretation of the data requires a considerable amount of background information in order to provide a useful service. It is, therefore, considered that the biological methods offer the best approach and further investigations should be carried out on the aerobic and anaerobic incubation techniques. The chemical method of Jenkinson should also be investigated further as it appears to be complementary to the biological approach.

Discussion of the reports of syndicates 1A and 1B

Dr. D. S. Jenkinson commented on the use of chemical methods for testing soil nitrogen availability. He thought these were very much in the exploratory stage and should be thoroughly tried out in practice. He also said that there were a number of other chemical methods which had not been mentioned at this Conference and which would be worthy of consideration.

The idea behind the barium hydroxide method, on which he had been working was to attempt to get a measure of the labile organic matter in the soil. Barium hydroxide kept the humic part of the organic matter out of solution because it is electrically charged. Dealing with the analytical difficulties of the barium hydroxide method, he thought the use of concentrated sulphuric acid might be difficult in a routine laboratory. The precipitation of barium carbonate by atmospheric carbon dioxide was not important. The main difficulty with the method was that in working with very small amounts of glucose, small amounts of impurities such as cellulose might seriously affect the results. He thought that much more work should be done on these chemical methods and that perhaps they should be tried out first in the greenhouse.

Question 2

Which environmental and historical factors (in order of importance and magnitude) should be taken into account when predicting nitrogen requirements?

Report—Syndicate 2A

Chairman: Mr. T. H. Caldwell

Spokesman: Mr. D. Hewgill

The problem was dealt with from an advisory point of view; for the guidance of N.A.A.S. District Agricultural Advisers. It was assumed that advice would be given dealing with a particular soil type and under climatic conditions where the average optimum nitrogen requirement was known for most crops. This information would usually be derived from field experimental work and advisory experience over the years.

The 'average' nitrogen requirement for a particular crop should be modified according to special conditions prevailing on the farm; the 'average' nitrogen requirement would be dependent on soil type and would differ greatly on two extreme soils such as a Fen peat or Yorkshire sand.

Dealing with amending factors, necessitated splitting farming types into arable farms (including vegetable holdings), grassland, fruit and glasshouse.

Arable and vegetable systems

1. The most important factor was considered to be the previous history of the field. This includes:

(a) The influence of crop rotation, where a ley/cereal rotation may be compared with a continuous cereal system; the latter often requiring much greater amounts of applied nitrogen.

(b) The influence of previous manuring, as there may be a residual effect due to nitrogen applied in the previous year. More important, however, is the use of farmyard manure where an allowance of up to 30 units N per 10 tons may be necessary.

2. The system of growing was considered to be the second most important factor. In this respect, placement of nitrogenous fertilizers can increase the efficiency of utilization and reduce the average crop requirement. Irrigation may have a considerable effect on crop growth, but the adjustments in nitrogen requirement are still somewhat obscure. Spacing distances of the crop may affect nitrogen requirements and the use of 4-in. drills for cereals needs consideration.

3. Soil condition was considered to be the third most important factor as poor tilths often increase the crops' need for nitrogen.

4. The interactions of phosphate and potash with nitrogen were considered to be important because on highly productive soils, high levels (of phosphate in particular) may allow more nitrogen to be used effectively.

5. A factor which was difficult to place in the table is that of disease because in some cases where there is a previous history of disease this can have a major effect on crop growth. 'Take All' in cereals can be quoted as an example of a disease which may result in higher nitrogen levels being required to maintain production.

6. Of less importance, generally, are the following factors:

(a) Loss of nitrogen by denitrification.

(b) Gaseous losses of nitrogen.

(c) Ammonium fixation.

(d) Additions of nitrogen from rainfall.

These are all more difficult to predict but fortunately they rarely occur except under special conditions.

Grassland Systems

In these systems the optimal level of nitrogen depends largely on the intensity of farming and the means of utilizing the grass. Some systems rely mainly on clover for nitrogen production whereas others use fertilizer nitrogen according to grass requirements. In the latter case a linear response to nitrogen is usually assumed, up to quite high levels of nitrogen application.

Soil type was considered to be generally less important than with arable systems but certain climatic factors greatly affect nitrogen requirements.

1. Under a given set of soil and climatic conditions the system of farming was considered to be the most important factor.

2. The grazing animal was thought to be important because of the recirculation of nitrogen to the soil. In this respect an increase of stock numbers can increase the intensity of production. Alternatively, the use of nitrogenous fertilizers may increase grass production and release more land for arable cropping.

3. The 'timing' of nitrogen applications can be used to extend the growing season and increase production during the season. In grassland systems this was considered to be one of the more important factors.

4. In drier conditions, production can be increased by irrigation, less nitrogen being needed to maintain the same herbage output.

5. Many of the factors affecting arable crops (e.g., soil condition, interactions with phosphate and potash and disease), may affect the nitrogen requirement of grassland but it was not possible to place these in order of importance.

Fruit growing systems

Much less information is available on the nitrogen requirement of fruit crops. For apples, however, one major factor is whether the orchard is grassed down or clean cultivated. The former initially requires more nitrogen than the latter, but after a few years the difference may be reduced.

Glasshouse systems

The total seasonal nitrogen requirement is known for certain glasshouse crops, e.g., tomatoes, chrysanthemums and carnations. These requirements must be adjusted according to the system of growing and take into account factors such as steam sterilization, flooding, and the use of bulky organic materials such as farmyard manure, peat or straw.

Additionally most of the factors applying to the arable systems also apply to glasshouses.

In all four systems reviewed it is apparent that there are wide gaps in the Soil Scientist's knowledge, although for many factors some adjustment can be made.

Report—Syndicate 2B

Chairman: Mr. R. Thomas

Spokesman: Mr. G. R. Davies

The Syndicate found it impossible to give a set order of importance, but rather that the order would be varied according to the situation. Priority is complicated by the inter-relationship of one factor with another. However, some features are considered to be important in all cases. Special crops or circumstances would demand assessment of additional factors.

1. The most important consideration is the intensity at which the farmer is capable of farming. This imposes a limit on nitrogen use.

2. In the field it is necessary to study any possible limiting factors to optimal crop production, e.g., degree of acidity, drainage conditions, standard of cultivations, disease and pest incidence and the general fertility status of the soil. Any gross variations within the field would also need to be noted.

3. The previous history of the field is of prime importance. Details of crop rotation, period of time since last under grass and/or a leguminous crop and the manurial programme carried out for the past few years must be considered.

4. A knowledge of the nitrogen status of the soil by an analytical technique, (if practicable) is desirable. Directly related to obtaining knowledge of the nitrogen status is the texture of the soil; depth of the soil available for root development; amount, type and distribution of organic matter present and rainfall during the past winter.

5. The variety of the crop to be grown should be known together with information on the method of application of fertilizer normally carried out by the farmer. Similarly the probable water regime under which the crop will be grown, i.e., average climatic conditions and the possibility of irrigation must be taken into account.

6. Soil condition and the condition of the last crop can be of importance.

7. In the case of grassland the utilization of the grass is important, i.e. intensity of the enterprise and whether the herbage is to be cut for hay or silage or simply grazed. This would be dictated in part by the type of sward present and the legume/grass proportions.

8. In dealing with horticultural produce, the quality of the product must be also taken into account; for fruit production, nitrogen manuring would be dictated by the appearance of the crop and its past performance.

9. In all cases the advice must be economically sound.

Discussion of the reports of Syndicates 2A and 2B

Mr. L. J. Hooper commented that both syndicates appeared to have placed soil-water relationships low on the list of priorities for predicting response to nitrogen; he thought that they should be considered of high importance. *Mr. G. R. Davies* replied that his syndicate had not implied any order of priorities but thought it was not possible to place the important factors in any order of priority as they were all equally important.

Question 3

Which lines of research and investigational works appear most promising and how should these be pursued in the immediate future?

Report—Syndicate 3A

Chairman: Dr. R. C. Little

Spokesman: Mr. W. M. Waller

The syndicate was of the opinion that the question asked was very wide and therefore restricted its discussion to those aspects of research and investigational work which had advisory implications.

1. *Nitrogen metabolism*

(a) *In the soil*

More work was required on the reactions involved in the nitrogen cycle and on the products of those reactions. Aspects of this which were considered of particular importance were:

(i) The effect of partial sterilization on the nitrogen cycle and particularly its effect on the possible accumulation of both inorganic and organic toxins such as nitrite.

(ii) The effect of incorporating large quantities of peat into glasshouse soils.

(iii) The value of inhibitors in controlling metabolic processes for research and possibly agronomic purposes.

(b) *In the plant*

(i) The effect of the source of nitrogen on cation uptake and, its effect on the composition and the quality of the plant (including grain etc.).

(ii) The effect of light on nitrogen metabolism within the plant.

2. *Residues and losses of nitrogen from the soil*

Information was required on:

(i) The nature and distribution of fertilizer and crop residues in the soil profile. There was need for further study on the beneficial effects of deep rooted legumes, particularly to differentiate between the effect of nitrogen residues and any beneficial physical effect on the soil profile.

(ii) The effect of soil type and of climate on losses of nitrogen from the soil profile.

3. *Methods of estimating the nitrogen status of the soil*

It was appreciated that this topic was being discussed by another syndicate, but it was considered that more work was required on the methods of estimating the nitrogen supplying power of soils and on the application of these methods to the production of the amounts of mineral nitrogen released on steam sterilization.

4. *Nitrogenous fertilizers*

Further studies were required on:

(i) The effect of compounding on:

(a) The chemical form of the materials used in making compound fertilizers.

(b) Changes in the form of nitrogenous compounds in compound fertilizers after their application to the soil.

(ii) The utilization of the large amounts of N available in the form of poultry manure, slurry, etc.

5. *The use and utilization of nitrogenous fertilizers*

The Syndicate supported the observations of Mr. Dermott in his paper 'Nitrogen and Advisory Problems' that more information was required on:

(a) The relationship between nitrogen and soil structure.

(b) The relationship between nitrogen and water.

(c) The responsiveness of soil series and soil types to nitrogenous fertilizers, and the factors which lead to variations in response.

(d) Methods of obtaining higher efficiency from fertilizers such as urea.

(e) Methods of improving the efficiency of utilization of nitrogen by arable crops.

The Syndicate was asked to comment on how the work should be pursued; much of the work was in the province of research stations, but clearly it was the function of the advisory service to apply the findings of the research workers to practical agriculture and to prove that the results obtained on the specific soil types of the research stations were applicable to the wider range of soils found in the regions.

Report—Syndicate 3B

Chairman: Mr. G. H. A. Edwards

Spokesman: Dr. D. B. Davies

1. *Nitrogen Response*

Reasons for differences in nitrogen response require greater consideration. Variation in response caused by variation in the following factors should be studied further:

Water regime and irrigation
Soil profile, particularly structure
Temperature
Farming system
Plant root distribution
Pests and disease

N.B. In particular this means that the field trial work must be supported by more environmental assessments and that the design of trial work must be based on this necessity. Suitable statistical techniques should be applied wherever possible.

Field trial work should be more closely related with fundamental research work particularly in physiological and biochemical fields. For example, correlations of nitrate reductase activity in plant tissue with nitrogen responses should be made.

Implementation will involve close personal liaison between N.A.A.S. Soil Scientists and research workers, and where large programmes of work are involved the co-operation of the A.R.C. should be sought.

2. *Nitrogen and Quality*

Further work is required on the effect of nitrogen usage on crop quality. The syndicate had in mind the effect of:

(i) Type of nitrogen fertilizer.

(ii) Timing and levels of application.

In particular, the quality of grass in relation to animal health and grass utilization requires further investigation.

3. Further investigations into soil nitrogen reserves are required. The form in which they occur, their availability to plants and suitable laboratory methods for their estimation including leaf analysis should be pursued. It must not be forgotten that comparisons of mineralization of nitrogen in soils are pointless unless methods of sampling, drying, grinding, sieving, storing etc., are rigorously standardized.

4. *Grass and clover*

Variations over the country in efficiency of nitrogen fixation by clovers and of utilization of nitrogen by grasses should be determined.

5. *Nitrogen and horticulture*

(a) The N.A.A.S. should attempt to assess to what extent existing data on the nitrogen requirements of horticultural crops is generally applicable to commercial holdings. This refers particularly to the information for glasshouse crops and simple trials should be conducted to elucidate this.

(b) The lack of critical information on the nitrogen requirements of vegetable crops on a range of soil types should be remedied.

(c) The possible role of nitrogen transformations in 'sterilization injury' should be further examined and resolved.

Discussion of reports of Syndicates 3A and 3B

Dr. D. A. Boyd commented on the statistical correlation between results of field experiments and laboratory analyses on the soil. He said that refined statistical techniques were available and what was needed were field experiments of sufficient quality to use these techniques. He thought that future field experiments should be very well planned at an early stage.

Chairman's closing remarks

The Chairman said that it would be quite impossible to sum up, or attempt to sum up, the proceedings of the Conference but he would like to refer to one or two points. It seemed that there was still much information not being fully used by farmers and growers. There was also an urgent need for much more precise knowledge to deal with problems which exist at present and to anticipate problems likely to arise in the near future. Information was lacking on the precise use of nitrogen for many crops in varying circumstances. Unfortunately the methods of soil analysis for nitrogen whether chemical or biological, seem unlikely to be acceptable in connection with routine advisory work for some time to come. It may be that advisers were aiming at too great a degree of precision but in the short term it is only possible to use the results of experiments, properly interpreted, in conjunction with advisory experience. In the meantime, however, from carefully conducted experiments, he felt that advisory work still seemed to be a very personal matter between the adviser and farmer or grower and that a good adviser is invaluable, as so much depends on his interpretation of what information is available and on his ability to see, observe and interpret the facts of each particular case.

The Chairman thanked the speakers, all who contributed to the discussion and the organizing committee.

Author Index

Dd. 147004 K10 1/69 SBN 11 240915 6